Knowledge, Organizational Evolution, and Market Creation

For Ricky, Ajay, and Aditi

Knowledge, Organizational Evolution, and Market Creation

The Globalization of Indian Firms from Steel to Software

Gita Sud de Surie

Senior Fellow, Wharton School, University of Pennsylvania and Assistant Professor, School of Business, Adelphi University, USA

Edward Elgar

Cheltenham, UK • Northampton, MA, USA

Published by
Edward Elgar Publishing Limited
Glensanda House
Montpellier Parade
Cheltenham
Glos GL50 1UA
UK

Edward Elgar Publishing, Inc.
William Pratt House
9 Dewey Court
Northampton
Massachusetts 01060
USA

A catalogue record for this book
is available from the British Library

Library of Congress Control Number: 2008926581

ISBN 978 1 84720 456 1

Printed and bound in Great Britain by MPG Books Ltd, Bodmin, Cornwall

Contents

Figures

Tables

Preface

Gita Sud de Surie

The genesis of this book was a visit to India in August 1992, a visit that was momentous in shaping events in my life for the next ten years. I relocated to India to study Indian organizations for my doctoral work in 1993. After completing my doctoral work in 1996, I stayed on as an academic in India until the spring of 2001. This prolonged immersion in India, both as a researcher and academic, provided me with a unique advantage – a position from which I could be an insider. Consequently, I was able to experience first hand the structure of Indian organizations and industry linkages with other institutions, both locally and globally, and to study their evolution during the period after economic liberalization was initiated.

In this book, I aim to trace the evolution and internationalization of Indian firms over the last decade. I focus on the role of knowledge in organizational evolution and in economic development. In addition, I hope to capture the dynamism and spirit underlying these changes by adopting a lens that allows for multifaceted perspectives. Using organizational data from my field studies, I examine the nature of these microeconomic and organizational changes and their implications at the macro level. These studies suggest that these micro changes engendered an understanding and reinterpretation of national identity, aligning it with those prevailing in industrialized countries.

I am deeply indebted to the organizations that participated in the initial study and permitted subsequent rounds of interviews at various intervals since 1993, in 2003, and most recently in January 2007, and to all interviewees too numerous to name individually. I would like to thank especially all managing directors/organizational heads in India who made the studies possible, and to the Confederation of Indian Industries, the Consulate General of India in New York, and state industry ministries who helped facilitate the work.

I am grateful to Bruce Kogut, Jitendra Singh, Michael Useem, Howard Perlmutter and Larry Hirschhorn at the Wharton School of the University of Pennsylvania for helping to encourage, inspire, and evolve the initial study in 1993. From mid-2001, collaborations with Harbir Singh and Lori Rosenkopf and interactions with Dan Raff, Peter Cappelli, Mike Useem, and others at seminars and conferences at the Management Department at Wharton were invaluable. I thank Mark Zbaracki, Gerry McDermott, Saikat Chaudhari, and others for comments on my work. I thank the late Clifford Geertz of the Institute of Advanced Study in Princeton whose books and encouragement inspired me and to whom I owe the title of the final chapter. I also thank Shiv Visvanathan, S. Irfan Habib, Dhruv Raina, and other scientists in India for helping me to understand the state of science and technology in India from an academic perspective. Conversations with Bimal Jalan and R.A. Mashelkar also provided historical context and inspiration. I thank Jeff Goldstein and Jim Hazy at Adelphi University for helping me to develop a perspective on complexity theory and other colleagues at Adelphi for a continuing dialogue. I also thank Christine Shaw for giving me the historical background on economic development. Any errors and omissions are mine alone.

Finally, I thank my husband, Ricky, for his unflagging enthusiasm, encouragement and support, my parents-in-law Atam Dev and Vimla Nagrath Surie, sister-in-law Meena Wilson, and children Ajay and Aditi for all they have done for me. I thank my parents, the late Baldev Sud and Krishna Kanta Ahuja, for having instilled in me a love of learning and discovering how the world works. I thank my brother Sunil Sud and nephews Adrian, Jai and Nathan for help with the figures and tables at the prepublication stage and my sister-in-law Donna for creating a nurturing and productive work environment. Finally, I would like to thank all my friends both in India and in the United States for their unfailing support and good cheer.

Foreword

Bruce Kogut, Sanford C. Bernstein Professor of Leadership and Ethics, Columbia University

Ten years ago, it would have caused just about anyone to smile at the prediction that by 2007 the world steel industry would be led by an Indian entrepreneur who had taken over the prize gems of French, Luxembourg, and Spanish genius. However stunning, the acquisition of Arcelor by Lakshmi Mittal represented more a bold financial strategy than a signal of India's own technological prowess. Nevertheless, this acquisition was followed shortly after by the takeover of the Anglo-Dutch company Corus by Tata, which had slowly upgraded its technological capabilities in its Indian operations over decades. Clearly, it could not be denied: something had happened in India.

But what? It is tempting to look at software and information technology for clues, for India has undeniably generated several global powerhouses in this sector. Less than a decade ago, the Indian software industry was called a 'body shop' where boring code was produced according to Western designs. I recall a meeting that I organized at the Wharton School – where Gita took her PhD – of top Indian and American software managers to discuss the then mild topic of 'outsourcing'. A young head of the software outsourcing unit of a large American multinational was outright dismissive of the thought that Indian software firms would be more than a remedy for bad code (the 'Y2K' nightmare) and dependent party to software innovation. It was an interesting meeting because many of the Indian software managers in that room went on to lead their companies to the top ranks of world software companies by 2007.

The idea that Indian companies might soon be major players in pharmaceuticals seems today equally farfetched. Of course, many Indian companies are already important manufacturers of generic drugs. The innovative capability to compete in biotechnology seems today a distant objective. Acknowledging all the caveats in logic by analogy, is it nevertheless not reasonable to ask if the experiences of the steel and software industries might at least demand a more cautious assessment of the speed of technological learning in some of the poorest countries in the world?

The world, so unfairly marred by an unequal distribution of income and wealth, is most cruel in wasting the most precious resource of human minds. No wonder we often see many of those who have the dual good fortune of being bright and educated leave their countries for want of opportunity and make their careers in more wealthy countries. If only these minds would return to their countries could we then see real progress, it was often believed.

But this belief, which so often plagued the conscience of those who did leave, is false. For the great depository of collective intelligence is not the aggregation of individual brains, but their organization. It is the organization that harnesses their capabilities, provides the incentives for innovation, and gives the capabilities by which to achieve and implement new technological knowledge. The sudden emergence of Indian multinationals is not simply a story of individual intelligence and leadership – though one cannot doubt their importance. Rather, this emergence is the accretion of decades of collective learning within Indian firms, which, when released from the shackles of a too paternalistic government, entered quickly into the world scene as multinational companies.

Gita Surie has gathered in this book a priceless account of the 'micro histories' of some of the most prominent Indian companies in the industries of steel, software, and biotechnology. The history of the steel companies focuses on their evolution from dependence to innovators by the creation of a rich community of practice that could absorb and improve the importation of foreign knowledge. Software is not the same story. The firms are often relatively young and they were not created in reference to national industrial planning. Their dynamic is more condensed and their rise more rapid. The biotechnology industry is the darling of the trio. Here Gita Surie emphasizes the experimentation in organizational designs and the implementation of a learning heuristic that relies upon incremental expansion.

What happened then in India? Gita Surie's answer is that India developed over recent years highly capable firms that progressed by organizational learning and evolution. Of course, in the background are such factors as the impressive technical institutes, the cost advantage for employing engineers, and the favorable tax treatment of exports. These factors have enabled Indian firms to expand, but they do not explain their success.

Ultimately, the story that Gita Surie offers us is an inside look at the historical development of organizations in an emerging market. The detailed analysis shows a history of trials, many of which failed, some that succeeded. Economic progress is more than the addition of so much labor and so much capital. It is rather the outcome of the organization of these factors of production.

It is the growing ability of some Indian firms to organize better than firms located in other countries that explains their entry into the ranks of the largest multinational corporations. This may seem as a sudden emergence from the outside. Inside, it is the outcome of a lot of hard work and learning. If the book holds a general lesson for development, it is that organizations, and not only factors of production or institutions, matter and matter a lot.

1. Introduction

This book documents the journey embarked on by firms in five different industries over the period of a decade (1993–2003) beginning shortly after the liberalization of the Indian economy in 1991. It includes 'old economy' industries such as steel, automotive components, heavy equipment and other manufacturing industries, and 'new economy' industries such as software and biotechnology.[1]

Internationalization has largely been examined from the perspective of industrialized countries with some exceptions (Dahlman and Westphal, 1982; Westphal, Kim, and Dahlman, 1985; Lall, 1987; Enos and Park, 1988; Chang, 2003; Guillen, 2005) of studies on how developing countries and emerging economies build capabilities. Although problems and issues related to development have been widely studied, the emphasis has been on technology transfer by foreign firms and its attendant difficulties rather than on the innovative capacities of recipient firms. This book aims to redress this balance by adopting an evolutionary approach to capability-building and focusing on innovation in India, a developing country, and to situate this process in the context of a transition to a market-oriented economy. Although the perspectives applied in this book are not new, they have been synthesized and applied in a novel context – an emerging market.

CONTEXT

I draw on a variety of disciplinary perspectives to develop a theoretical framework for the creation of organizational and technological capabilities. It combines perspectives on social learning, internationalization, knowledge and capabilities, real options, and complexity theories to yield a deeper understanding of *how* internationalization occurs. The analysis focuses primarily on firms and industries, while the institutional context and regulatory environment provide the backdrop. I begin with a micro-level examination of the firm as the context for learning and innovation via the adoption of new technologies and knowledge from external sources. Organizational and technological skills are developed by adapting these new technologies for local use. Successful adoption and assimilation of new technology, its diffusion and innovation are dependent on the creation of communities of practice that encompass inter-firm relationships with foreign suppliers of technology, inter-departmental relationships within the firm and relationships with local suppliers. I examine this process of community creation via an in-depth study of three firms in the steel, construction equipment, and bearings industries located in Eastern India. These case studies are corroborated using supplemental data from seven other firms located in Northern, Western, and Southern India. Findings suggest that internationalization begins with the adoption and replication of technology and associated organizational practices and culminates in an identity shift and aspiration adjustment towards becoming 'world class'.

Despite improvements and capability-building in the manufacturing sector during the initial stages, global oversupply, commoditization, and lack of economies of scale constrained the emergence of Indian firms as important players in global markets. Nevertheless, these early attempts to modernize and globalize initiated a wider dissemination of the methodological approach of scientific discovery and triggered an introspective examination by firms to discover 'core competences' and 'inimitable resources'.[2] In subsequent stages, Indian manufacturing firms took a more proactive stance in initiating internationalization using an arsenal of techniques acquired as a result of their encounters with foreign firms.

INDIAN SOFTWARE

In contrast, the emergence of a new industry, software, catapulted India onto the global stage. Indian participation in software was not hampered by the regulatory environment, which focused, instead, on computer hardware. While the computer industry was designated a strategic industry, under the regulatory regime of the 1970s and 1980s, it suffered the same regulatory restrictions and requirement of indigenization imposed on other industries. Consequently, the software industry was a peripheral offshoot, largely unregulated in the 1980s with small firms emerging to provide services in the vacuum created by the withdrawal of IBM from India in the mid-1970s (Heeks, 1996).

Moreover, software was an emerging industry worldwide; leading firms like SAP, Microsoft, Sun Microsystems and Oracle were established, starting in 1972. Indian entrepreneurs with technical skills acquired at educational institutions like the Indian Institutes of Technology seized the opportunities created by the emergence of a new industry and the presence of a critical mass of skills available in India. Moreover, Indian entry into foreign markets was facilitated by rising demand for low-cost computer programming. The large pool of programmers available at lower wages presented a labor cost advantage that Indian firms were able to leverage for their own growth during the 1990s. Advances in technology and experiments with the Internet for commercial use also made it possible for firms in developing economies to participate in cross-border work leading to its reorganization and spatial distribution. Just as manufacturing firms in the industrialized world had earlier sought flexibility through access to manufacturing locations in developing countries, knowledge and technology-intensive firms in the 1990s began to seek low-cost knowledge inputs from developing countries at earlier stages of the technological life cycle to alleviate uncertainty and speed the introduction of new technologies. Consequently, the emergence of knowledge-based industries in the developing world and its trajectory complements its increasing maturity in the industrialized world.

The global success of Indian software also sensitized firms in industrialized countries to the advantages of tapping India's technically skilled workforce in other knowledge-intensive domains such as biotechnology. While government sponsorship of science and technology and incentives to build this industry rendered it attractive for local firms to invest, foreign multinational corporations (MNCs) also began attempts to reduce the cost of innovation while establishing a foothold in new markets. I present evidence from both biotechnology and software firms and outline the evolution of participation in the global economy by both Indian and foreign firms in these industries to contrast with the evolution of manufacturing firms.

RESEARCH QUESTIONS

This book seeks to answer the following questions.: How do firms in developing countries grow and expand across national boundaries and what capabilities enable some firms to outperform others consistently? What organizational processes and practices are effective in enabling participation in innovation and in fostering wealth creation in an emerging economy? What factors are impediments, and what ideological shifts are required for transformation? I do not judge local culture or values. However, I contrast the new ideologies of globalization with earlier cultural assumptions to show how adopting new cultural values sets developing societies on the path of transformation and generates new options.

This book's central thesis is that firms evolve and grow by developing internal capabilities and adapting to changes in the environment. Learning and innovation are the key mechanisms underlying international expansion and occur through interactions with domestic and international firms and national and international institutions.

ORGANIZATION OF THE BOOK

The second chapter outlines theories used in examining firms in both old and new economy industries. It combines evolutionary and knowledge-based approaches to understand the emergence and growth of firms in these industries and study how innovation occurs. Evolution occurs through knowledge replication and the ability to access new knowledge. In the absence of knowledge in the local environment, it must be sought externally. The evolution of new capabilities in domestic firms transforming them into multinationals involves several steps:

1. In the first stage, community creation is critical to enable knowledge transfer.
2. In the second stage, which is transitional, newly acquired capabilities are institutionalized.
3. In the third stage, firms must scale operations, develop complementary capabilities and access new markets and financial resources to gain membership and centrality in the global community of firms.[3]

However, the environment must be sufficiently diversified to absorb the output of knowledge creation activities. Firm growth is also limited or inhibited by aspiration levels, the level of specialization in the overall economy and the ability of firms to accommodate heightened complexity. Nevertheless, to overcome the limitation of low specialization in the local economy, the community need not be confined locally but can be globally distributed. Likewise, the pursuit of deliberate organizational change can enhance the capacity to absorb complexity.

This framework yields insights on internationalization by developing country firms that may be applicable, more generally, for the strategic management of organizations. While strategic management literature has traditionally emphasized rationality and efficiency in decision-making, this book suggests that knowledge transmission requires the presence of social communities. However, the paradox is that while knowledge transmission necessitates enhancing boundary-crossing interactions to accelerate innovation, appropriating knowledge requires that inter-organizational boundaries must be defined and constructed. Thus, by examining cross-border participation in innovation and inter-firm boundaries, this book also touches on the question of the boundary between the firm and the market and, consequently, contributes to literature on the theory of the firm.

Chapter 3 presents the research methods used and provides details about the evolving research process. The research uses a multiple case-study design to examine the factors influencing shifts in organizational practices and processes that affect the boundaries of communities of practice in international innovation (Campbell and Stanley, 1966; Yin, [1989] 1994). Such a design is advantageous when randomization of subjects is not possible, when relevant behaviors cannot be manipulated, and when examining phenomena within a real-life context (Yin [1989] 1994). Data are drawn from over 145 in-depth interviews, observations in situ (meetings, workshops, presentations), from company annual reports and industry publications, and conferences hosted by both academia and industry.

Chapters 4 and 5 document the experience of firms in the manufacturing sector and their evolution into firms that participate in the global manufacturing arena. The first two stages of their evolution are documented in Chapter 4. Learning is both adaptive and evolutionary, leading them to become members of a global community of manufacturing firms. The adoption and replication of practices and processes, and interactions with members of the community allow them to become participants, albeit peripherally. This is an iterative process involving adaptation of behavior, dialogue, and new consciousness of what global membership entails.

Chapter 5 highlights the third stage of global expansion and integration in manufacturing firms. It suggests that, at this stage, successful technology adoptions in these firms led to the internal recognition that growth and expansion would require new strategies to enable participation in a market that was increasingly complex and no longer confined by national boundaries. Thus, new strategies and rules were adopted as moves to position these firms in international markets; in addition, organizational design had to be altered to match strategy to deal with heightened complexity.

Chapter 6 contrasts the experience of manufacturing firms with that of firms in high-technology industries. It traces the emergence and evolution of software and biotechnology firms that were global from the start and have less administrative heritage. Evidence suggests that strategy is dependent on pragmatism and that commitment to developing expertise is critical, especially when specialized expertise is lacking. The entry and integration of Indian firms into the global arena occurs via the adoption of a variety of roles – as suppliers, as partners in alliances, and as competitors of multinational companies. These roles provide a mechanism for integrating the knowledge of the community and enable the replication of the context of biotechnology and software in a new geographic location. Linkages with the United States and other locations are noted, including alliances and the movement of experts between these locations and India. Moreover, the codification of knowledge enables firms to industrialize the process of knowledge creation and develop and trade knowledge components that are used as real options in the expansion process.

In addition, in this third phase, interactions with foreign suppliers and buyers occur on a more equal footing, with greater emphasis on knowledge production and innovation. Since the environment must be sufficiently diversified to absorb the output of knowledge creation activities, at this stage there is increasing specialization in the activities of firms. Moreover, there is wider recognition that aspiration levels influence firm growth and that the level of specialization in the economy is not necessarily fixed. Consequently, this stage witnesses the rise of conscious entrepreneurial attempts by firms in developing economies to learn and grow internationally. Although specialization may be positively related to the level of growth since it requires economies of scale for global competitiveness, entrepreneurs in developing countries can aspire to become global players by using a globally distributed community rather than by remaining within national borders. A consequence of cross-border interactions is the diffusion of global practices and the creation of markets in new sectors of the economy.

The concluding chapter synthesizes evidence and provides a framework for the dynamics of internationalization and the emergence of new multinationals as a result of competition and innovation induced by multinational entry. International expansion through acquisition, subsidiaries and the ability to source work from the United States suggest that India may be on the threshold of a major leap forward. An implication is that development may follow a unique path and 'late movers' are not necessarily disadvantaged. A separate research study on biotech and software firms in the United States (Surie, 2004) provides additional insights on multinational firms and indicates an altered role for India in the international network for knowledge. I outline a new hybrid model of organization based on the Indian experience that is relevant for the requirements of societal development. I also compare India's development with that of other emerging economies like Korea, China, and Brazil, and conclude by summarizing a theory of firm growth across borders, suggesting a new interpretation of how 'national competitiveness' and 'country capabilities' are likely to evolve in the 21st century.

INDUSTRY BACKGROUND

This research was conducted in several phases. In the first phase of the research, I examined technology transfer in firms in manufacturing industries. This field study included an examination of three manufacturing firms in the steel, construction equipment, and automotive ancillary industries (such as bearings and pistons and rings) during 1993–96. Follow-up interviews were conducted on three of these firms in 2000 and 2003. The steel, automotive, and related industries are significant in the manufacturing sector. Driven by steel-intensive economic activity in many developing economies, global apparent consumption of steel increased on average by more than 7 percent per annum since 2002 to reach 1.113 billion tonnes last year. To meet this increase in demand, steel production accelerated sharply, reaching 1.24 billion tonnes in 2006, up by 393 million tonnes or 46 percent compared to its level of 850 million tonnes in 2001. China accounts for about 32 percent of the world's apparent steel consumption. Crude steel production in China rose to 423 million tonnes in 2006 accounting for 34 per cent of world production. In India, the world's seventh largest producer of steel, production reached 44 million tonnes. Russian steel production grew from 59 million tonnes in 2001 to 71 million tonnes in 2006. The share of other Asian countries (excluding China), NAFTA and the EU-25 has declined. Crude steel production in 2006 increased from 90 million tonnes in 2001 to about 99 million tonnes in the US; Japanese production reached 116 million tonnes and production in the EU-25 rose to 198.5 million tonnes in 2006 (OECD, 2007). Projected automotive capacity (2002–08) is expected to be the highest in Asia (excluding Japan; Hughes-Cromwick, 2003). Employment in the motor vehicle and equipment manufacturing industry is expected to increase 9 percent over the 2000–10 period (US Department of Labor, 2000–03). Forecasts for earthmoving equipment growth rates in 2005 (over 2004) range from about 5.6 percent worldwide, 7.8 percent in the United States, and 6.9 percent in Canada, a substantial acceleration from the rate of increase during the 1990s, driven by the ongoing economic recovery and expansion in the developing nations of Asia/Pacific and Latin America and favorable prospects of economic growth in Eastern Europe (Association for Equipment Manufacturers, 2004–05). US exports of construction equipment increased by 30 percent in 2004 compared with the previous year, an increase that represented US$8.9 billion[4] in equipment sold worldwide, with all regions showing double digit increases (Equipmentworld Magazine, 2005). Consequently, studying these industries is critical, particularly since emerging markets represent a large proportion of future demand.

The field study of the software and biotechnology firms was conducted during 2003. This book focuses on innovation in contexts in which tacit knowledge and experiential learning are important. Past studies suggest that biotechnology (Pisano, 2000) and software afford such a context. During 1993–99, the total R&D expenditure of publicly traded genomics firms grew 48 percent per annum (OECD, 2001). From 1996–99, US biotechnology trade grew by 13.2 percent a year on average, whereas technology transactions increased by 9.5 percent and total trade by 6.5 percent. The United States is a net exporter of biotechnology products and remains a leader on the international market. The share of biotechnology in the technology trade surplus is twice as large as its share in technology trade (0.9 percent), suggesting a US trade specialization (ibid.).

Similarly, the software industry, which is part of the information and communications technologies (ICT) sector, accounts for a large and growing share of investment and contributed significantly to output growth, particularly in the United States, Australia, and Finland in the late 1990s (OECD, 2002). Software firms are the most R&D-intensive of ICT firms, important recipients of venture capital (up to 20 percent of total technology venture capital in the United States, over 30 percent in Europe) and increasingly active in patenting. In the United States, software-related patents now account for between 4 percent and 10 percent of all patents, depending on how they are counted (ibid.).

Both industries are also rapidly gaining importance in India. The Indian IT services industry, which began in the mid-1970s, reached the US$10 billion mark in 2002 and was targeted to reach US$77 billion by 2008 (NASSCOM-McKinsey report, 2002). The bulk of the growth was driven by exports, which grew from US$2.6 billion to US$7.8 billion at a growth rate of 43 percent per year from 2000 to 2002 (ibid.). Worldwide spending on IT was estimated to have grown from US$1384 billion in 2004 to over US$1479 billion in 2005, a growth of nearly 7 percent over the year. Services, comprising IT services, product engineering and business process outsourcing (BPO) account for a dominant share (approximately 58 percent) of worldwide aggregate spend and form the fastest-growing segment (growing by 8 percent in 2005; NASSCOM, 2006). Indian IT exports grew from US$13.3 billion in 2003–04 to US$18.2 billion in 2004–05. Software and services exports are projected to grow at 32 percent in the current fiscal year (ibid.).

Similarly, biotechnology is an emerging industry in India. It accounted for just 2 percent of the global biotechnology market in 2003 and is estimated to grow exponentially over the next five years, with an expected global market share of 10 percent. The first Indian biotechnology company was established in 1978; currently there are over 150 biotechnology companies. Biotechnology is considered to be the next major driver of growth and the Government of India (GOI) has taken special initiatives to promote India's biotech industry. In addition, the biotech research plan outlay has been doubled from INR6.22[5] billion in the Ninth Plan (1997–2002) to INR14.5 billion in the Tenth Plan (2002–07). Apart from funding, the GOI has eased the regulatory framework by approving genetically modified crops, recombinant-DNA products (rDNA) and ethical stem cell research (ICFAI, 2005).

AREAS OF CONTRIBUTION

This book contributes to a deeper understanding of processes underlying internationalization and highlights the relevance of interactions between industrialized and developing countries in contributing to building capabilities and competitive advantage. The introduction of new technologies and innovations from external sources acts as a catalyst for new action in a traditional context, raising local aspirations and catapulting the firms with the strongest capabilities to the next stage of evolution. From an evolutionary perspective, exposure to new ideas alters the behavior of the system just as a problem-solving organism changes its behavior by permitting invalid assumptions to be discarded (Popper, 1963). The book thus suggests that behavioral changes are accompanied by changes in cognitive assumptions through action and links micro-level behavior with macro outcomes. In this respect, the book contributes to an understanding of the mind–body problem of philosophy in organizational and socioeconomic contexts. It also highlights the implications for transforming societies when changes in organizational practices usher in new cultural values.

The book also proposes a different theory of decision-making than one suggested by classical economics, by emphasizing that technology selection and decisions to innovate or participate in markets result as much from heuristics of aspiration adjustment as from rational search and optimization. Moreover, using bounded rationality heuristics rather than an optimization view of decision-making contributes to the robustness of the new system rather than to its failure because the resulting externalities yield societal benefits.

Finally, the book contributes insights on the theory of the firm by showing how the boundaries of the firm are extended through the creation and replication of communities of practice across borders. Such communities facilitate the generation and diffusion of knowledge both in and across firms. Knowledge transfer, absorption, and creation lead to the emergence of new identities and involvement in activities that speed the internationalization of firms in growing industries. Entrepreneurial firms source new knowledge and leverage capabilities through boundary-crossing interactions with other organizations (e.g., firms and institutions). The expansion of the boundaries of the firm via strategic alliances, joint ventures, and subsidiaries across borders also mirrors a restructuring or contraction in other firms and industries. Thus, it provides insights on the processes underlying shifts across industries through strategic action by firms in a developing economy.

NOTES

1. The book draws on studies conducted during this period (Surie, 1996; Surie and Singh, 2004).
2. The resource-based literature discusses resources in terms of their contribution to a firm's ability to use them in ways that make it difficult for competitors to imitate its products or services (see Dierickx and Cool, 1989; Barney, 1991). Resources can include core competencies that provide firms with similar advantages (Prahalad and Hamel, 1990).
3. These stages are used to communicate differences between levels of learning and change. A staged approach to evolution does not imply that progress is inevitable, neither does it preclude regress. Thus, growth and exit are equally possible.
4. Billion = one thousand million.
5. Indian rupees.

2. Stages of globalization: from knowledge transfer to industrialized innovation

This chapter presents a framework for understanding how firms in emerging economies develop and sustain new capabilities and explaining why some firms outperform others. It outlines the main thesis of this book that the international expansion of firms follows an evolutionary process and is an outcome of the development of new capabilities. The chapter discusses the nature and role of knowledge[1] in creating capabilities, three distinct stages in the evolution of capabilities, and their accompanying modes of organization (see Figure 2.1). It concludes by suggesting how cross-border replication of knowledge can be accelerated. By synthesizing different approaches to internationalization[2] and drawing on a variety of theories of organization such as evolutionary and dynamic capabilities perspectives, real options theory, learning theory, and complexity theory it is possible to discern patterns that indicate the presence of distinct stages and organizational configurations as the firm globalizes.

The delineation of capabilities into distinct stages is used to explicate a complex phenomenon: namely, the emergence and globalization of firms from a location designated as 'developing' instead of 'at the technological frontier' (Lall, 1987), rather than to suggest rigid or time-bound adherence to evolutionary processes.[3] In the early stages of economic development, which is designated Stage I, capabilities are lacking in the domestic environment.[4] Modernization requires investment in new industrial technologies in multiple sectors, yielding increasing returns (Arthur, 1989; Murphy, Schleifer and Vishny, 1989), and skilled labor to transform the economy by shifting the surplus from agriculture, the strongest sector, to technology and infrastructure (Rosenstein-Rodan, 1943, 1944; Lewis, 1970; Murphy et al., 1989; Nelson and Pack, 1999).

Investments in industries such as steel, textiles, automobiles, trucks, heavy manufacturing equipment, and computers aim at building self-sufficiency and local capabilities (Lall, 1987; Enos and Park, 1988; Heeks, 1996). However, domestic firms face challenges such as scarcity of capital, lack of technology, skilled workers and managerial capabilities as well as the political risk of policy changes.[5] Since, at this stage, firms are constrained to acquire new knowledge from external sources (Dahlman and Westphal, 1986), Stage I is referred to as the *knowledge transfer stage*.[6] In addition, transferring knowledge to local firms by adopting and assimilating new technology from foreign firms also necessitates adopting new organizational structures.[7] Moreover, organizational evolution is accompanied by cognitive change, inducing a shift in identity and aspirations and reducing cultural and technological distance between domestic and foreign firms.

During Stage II, the acquired knowledge and capabilities are diffused in the local environment by institutionalizing routines associated with the acquisition of new technology. Stage II is an extension of Stage I, a transition period during which other local firms also adopt these routines and begin to integrate them throughout the organization in preparation for the next stage of growth.

Stage III focuses on domestic firms' efforts to participate in global markets on the basis of their own capabilities by accelerating innovation and industrializing knowledge production in the local environment. Lall (1987), Enos and Park (1988), and Scott-Kemmis and Bell (1988) suggest that the penultimate stage in the acquisition of capabilities by local firms is the ability to generate new knowledge and innovations independently. A key challenge for firm survival at this stage is the ability to expand rapidly and scale operations. Consequently, this stage is marked by the search for methods and experiments to accelerate innovation by industrializing knowledge production. By pursuing particular technological trajectories, the most knowledge-intensive local firms become innovators and creators of knowledge; their interactions with foreign firms and independent efforts to expand and compete lead them to participate in the global knowledge economy.[8]

The next section describes learning processes. The rest of the chapter explains how, at each stage, facilitating learning necessitates corresponding organizational transformation to enhance Indian firms' capabilities. Each stage is also distinguished by the use of different modes of knowledge creation. Figure 2.1 presents the evolutionary model of knowledge transfer, innovation and internalization

LEARNING PROCESSES

The learning processes explored and developed here are based on Popper's (1994) premise that the growth of knowledge is a consequence of problem-solving. He notes that problems trigger a new aim in organisms, giving rise

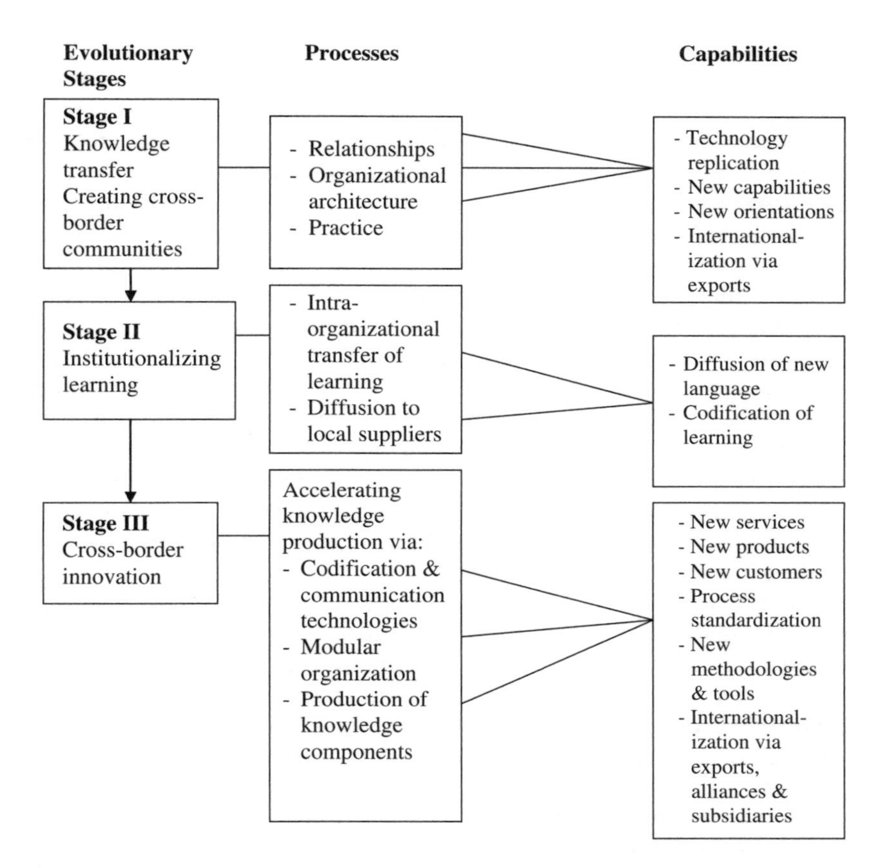

Figure 2.1 Evolutionary model of knowledge transfer, innovation, and internationalization

to solutions and novel behavior. Such problem-solving involves learning and innovation. Analogously, major changes in the external environment of firms in emerging economies trigger new problems, aims, and a search for solutions. In searching for solutions, a useful heuristic is to emulate high-performing firms.[9] Consequently, when faced with competition in a newly liberalized economic regime, Indian firms lacking technological capabilities to compete with firms from industrialized markets, engaged in problem-solving and sought new knowledge and technology from external sources to enhance competitiveness and survival chances. These attempts to transfer and absorb knowledge to create new capabilities from foreign sources involved a process of social learning, innovation, and legitimation[10] in the local environment.

Learning is viewed as practice-based and context-dependent rather than mere mechanical information transfer isolated from practice.[11] It involves a process of social construction and occurs in context via involvement in work, and is, therefore, 'situated' within a community of practice. Through apprenticeship with experts in the community, learners gain mastery of tasks, forge new identities, and negotiate meaning through socialization and membership (Brown and Duguid, 1991; Lave and Wenger, 1991, pp. 47–58; Wenger, 1998). Through such apprenticeship and participation in the tasks of the community, novices acquire mastery and adopt a new identity, that of a member of the community. Thus, the expertise and worldview of a particular community is transferred to newcomers.

By analogy, the adoption of new technology from external sources involves the creation of a cross-border community. The central task in transferring knowledge to build new capabilities is to facilitate knowledge absorption in a new context. Thus, in addition to transferring technical knowledge, firms must also be able to build a repertoire of new routines and transfer-related organizational capabilities. In short, this requires replicating the community.

Since practice is central to learning, developing cognitive and other capabilities requires that practice and participation are sustained through the creation of a community. Sustained participation and practice facilitate the development of two different types of knowledge: subjective knowledge and objective knowledge (Popper, 1994; see also Figure 2.2).

Subjective or personal knowledge grows through participation because active involvement in work shapes the personal and social experience of community membership. Also, participation allows individuals to negotiate and create meaning through their interactions with other community members. The psychological impact of participation in activity is to strengthen the new identity of the participant and provide a sense of affiliation and clues to appropriate behavior (Wenger, 1998).

In contrast, objective knowledge arises from attempts to solve problems through trials and experiments; solutions to an original problem usually lead to new problems (Popper, 1994). Such problem-solving is accomplished by studying and using artifacts, objects and tools that are products of the human mind rather than through direct social interaction. Examples of such artifacts include language, documents, tools, theories, rules, and procedures. Working with artifacts (for example, modifying a tool or a blueprint) facilitates problem-solving by enabling individuals to access the mental realm of creators of artifacts (ibid.). The act of creating new objects (e.g., producing a machine) enables individuals to externalize the mental process, engage actively and directly with the created object and make suitable alterations. Such efforts often yield new discoveries, solutions, and problems, sometimes as unintended consequences. The ability to externalize mental processes may yield results that often surpass what the individual may have thought he or she was capable of producing. As a result, concrete, observable, and lasting traces of the capabilities achieved are retained.

Thus, participation and substantiation (the creation of artifacts and objects) are two important interacting dimensions of practice. Participation involves the renegotiation of meaning in a new context. In contrast, by revealing new information, the creation of new artifacts makes possible the conditions for new meanings.[12] In turn, their interaction generates new knowledge that changes assumptions, provides new tools, theories, and perspectives. Therefore, facilitating learning and diffusing both personal and objective knowledge requires organizational environments that support both participation and substantiation.

STAGE I – KNOWLEDGE TRANSFER VIA COMMUNITIES OF PRACTICE

The ease or difficulty of knowledge transfer depends on the degree of tacitness and extent to which knowledge can be codified (Winter, 1987). Complex technical knowledge is often tacit and embedded within social systems. Consequently, its transfer and replication requires the formation of communities (Kogut and Zander, 1993). Inclusion in the community is critical since learning requires involvement in practice. Peripheral participation and legitimacy are forms of participation that are structured to make the practice of a community accessible to novices and non-members (Lave and Wenger, 1991). Thus, promoting learning requires that learners (participating peripherally) experience an approximation of full participation with access to full practice and communication (i.e., to computer mail, to formal and informal meetings, to telephone conversations, and to war stories, problems, and challenges faced by the community) and the know-how of experts (ibid., pp. 29–43). Moreover, they must be granted sufficient legitimacy so that the mistakes they make become an opportunity for learning rather than cause for dismissal, neglect, or exclusion. Denying learners legitimacy, access to information, or opportunities to practice makes learning difficult (Brown, Collins, and Duguid, 1989; Lave and Wenger, 1991).

Promoting learning, therefore, requires a modification of the information exchange system to enable new participants to perform tasks and solve problems effectively. Since existing information channels may be inadequate, alternative channels and feedback systems may have to be devised to ensure adequate bandwidth for problem-solving.[13] A consequence of attempts to establish new information channels may be resistance to change from members who previously controlled information and fear losing power.

Therefore, power and authority are issues that must be dealt with to ensure learning. The circulation and dissemination of new ideas also dictates acceptance of authority based on expertise by the community[14] besides authority based on personality or by virtue of bureaucratic position.[15]

Novices acquire expertise by participating in the community of practice (Brown et al., 1989; Lave and Wenger, 1991). Mastery and expertise involve the ability to participate consciously in the culture through its social network and language indicating the growth of subjective knowledge. It also involves the development of objective knowledge as evidenced by the ability to perform complex tasks, create new products, improvise solutions, and innovate, a natural consequence of day-to-day problem-solving within the community of practice.

At a collective level, practice enables the community to re-conceive its environment and identity. New perspectives of the world emerge from practice and from boundary-crossing interactions with other communities, helping to align a local regime of competence with other trajectories and linking the local with the global (Wenger, 1998). Key processes of this model are outlined in Figure 2.2.

Mode of Knowledge Production

When knowledge is tacit and contextually embedded, the holistic or craft mode[16] of knowledge transfer via the creation of a community of practice is necessary. Based on the learning processes outlined above, this involves:

1. enabling membership through relationship development to build trust;
2. aligning organizational design to facilitate boundary connections and alignment with other communities to enable information flow, problem-solving and innovation;
3. access to practice to promote the negotiation of meaning and identity.

The community of practice refers to all organizational sub-units directly involved in technology transfer and the foreign technology supplier. Extensions of the community include other intra-organizational sub-units, and the firm's local suppliers and customers.[17]

First, in cross-border knowledge transfers, community creation is essential for the development of local competence. Trust is widely recognized as contributing to cooperation, information exchange, and enhanced system efficiency (Arrow, 1974; Coleman, 1988) through reduced transaction costs

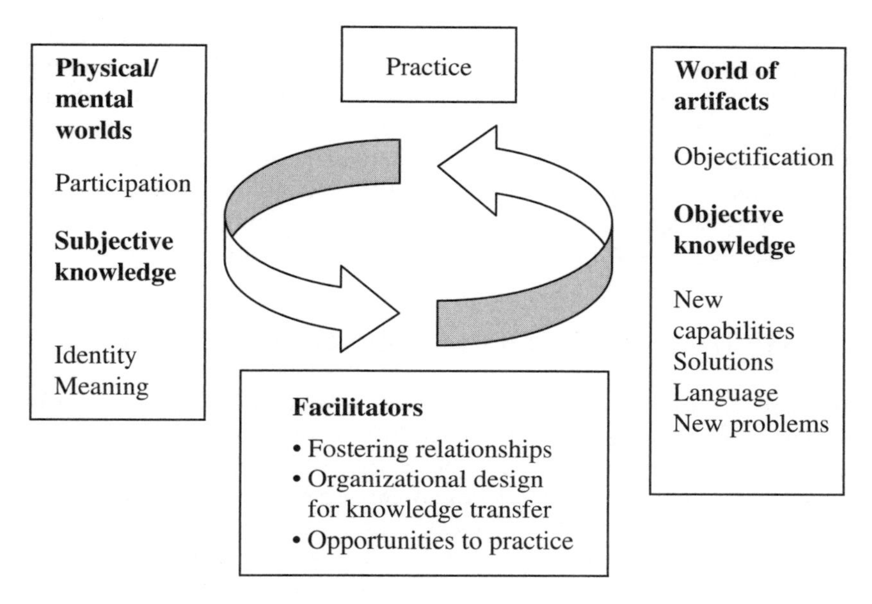

Figure 2.2 Learning via practice

(Kramer, 2000). Lack of trust hampers communication, collaboration, and involvement in the supply of training, access to knowledge (hardware, blueprints, and expertise) and impedes the provision of appropriate guidance and role models. It also impedes the creation of a common language to speed information transfer (Arrow, 1974). Finally, in the absence of effective partner relationships, active participation may not be achieved, thereby preventing the discard of entrenched assumptions, perspectives, and theories, and the acceptance of new ones.[18]

Second, organization design is critical in promoting the diffusion of subjective and objective knowledge through effective boundary-spanning mechanisms between internal departments, local suppliers, and other members of the community. Organization theorists have long observed that while specialization increases efficiency (Thompson, 1967; Lawrence and Lorsch, 1969; Galbraith, 1973), it also leads to different cognitive and emotional orientations (Lawrence and Lorsch, 1969) that impede knowledge-sharing. Thus, higher levels of integration and concurrent rather than sequential task processing are required for improving intra- and inter-organizational knowledge-sharing (Thompson, 1967; Clark and Fujimoto, 1991; Grant, 1996). Besides serving to disseminate knowledge, boundary-crossing mechanisms are also helpful in facilitating problem-solving and innovation.

Third, as noted above, opportunities to practice are essential for learning (Leonard-Barton, 1988; Brown and Duguid, 1991). If participation is aborted, inhibited, or delayed, learning to make new technology operative can be difficult, costly, and slow. Access to objects and artifacts helps hone new skills and perspectives and build a shared repertoire of routines across the community. Engaging with experts helps technology recipients create meaning and acquire subjective knowledge thus aligning the local community with the global community of manufacturing firms.

STAGE II – INSTITUTIONALIZING LEARNING

In Stage II, a transitional stage, knowledge acquired from direct participation in a technology transfer project is diffused to other parts of the organization not originally involved in the project. Transferring newly acquired concepts and practices involves developing a new organizational vocabulary, modifying existing routines and creating new ones until they are standardized, institutionalized, and taken for granted. This is usually accomplished by assigning project members who have mastered the new technology and requisite organizational changes to other areas of the organization. These new experts disseminate their knowledge by distilling and codifying what they have learned.[19] Institutionalizing learning (Figure 2.3) generates internal expertise and allows the recipient community to function autonomously.

STAGE III – FROM CRAFT PRODUCTION TO THE KNOWLEDGE ASSEMBLY LINE

Industrializing Knowledge Production

The craft mode of knowledge production and transfer involving face-to-face communication and collocation of communities is necessary in the face of difficulties in codifying tacit knowledge (Kogut and Zander, 1993; Surie, 1996; Surie and Singh, 2004). Replicating knowledge is a slow, difficult process in which competitive imitation takes time (Kogut and Zander, 1993). This is so because acquiring tacit knowledge requires that individuals must gain access to the social community, learn its 'codebook',[20] and master its worldview.

However, an alternative mode of transmitting and producing knowledge is feasible when technological advancements such as the Internet and new information and communication technologies (ICTs) bring previously

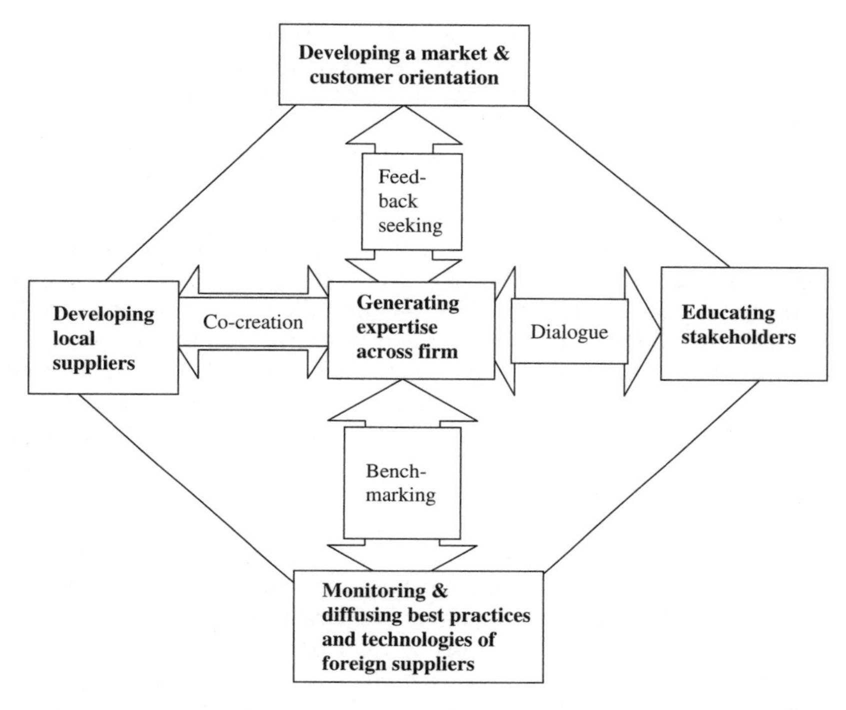

Figure 2.3 Institutionalizing learning by diffusing the community of practice

isolated worlds into contact. ICTs help to accelerate the transmission of knowledge by altering the emphasis on the two dimensions of practice – participation and substantiation – shifting the focus from the former to the latter. The use of ICTs permits the replacement of the craft mode of knowledge production, which relies on intensive participation, with an industrialized system relying on rapid knowledge codification. Moreover, the drive to accelerate knowledge production and innovation to compete effectively propels firms to adopt an industrialized system. Another important reason is the need to meet uncertainty by aligning strategy with organizational designs that incorporate flexibility through modularity[21] and enable rapid knowledge decomposition and recombination.

New communication technologies and software techniques (such as computer-assisted simulation and modeling) facilitate the industrialized knowledge production by reducing coordination costs and allowing rapid knowledge codification and dissemination. These innovations challenge the perception that knowledge codification is slow and difficult and that replication can occur only by forming epistemic communities (Cohendet and Steinmueller, 2000; Nightingale, 2000). Although some knowledge may be 'uncodifiable', the focus here is on knowledge that is, in principle, codifiable. The ease or difficulty of codification can also be viewed as a consequence of incentives (Nelson and Winter, 1982). Thus, what is tacit need not always be so; if economically viable, environments for codifying and exchanging knowledge can be constructed. Consequently, even newcomers with investments in learning and using these technologies who understand the 'codebook' are able to access knowledge. Although lack of contextual understanding is posited to inhibit the transfer of tacit knowledge (Brown and Duguid, 1991; Lave and Wenger, 1991), the problem may stem from lack of trust and a desire to withhold knowledge that is, in principle, codifiable, with those who are not perceived as members of the community (Surie, 1996; Surie and Singh, 2004). Thus, accessing the tacit dimension may not be an insurmountable task (Cohendet and Steinmueller, 2000; Cowan et al., 2000).

ICTs also permit more effective communication of 'symbolic' texts and, thus, promote the exchange of artifacts and tools to aid remote problem-solving. For example, by communicating scientific working models and observational data, a more fine-grained understanding of the underlying assumptions and concepts is possible than from summarized scientific results. Recipients can use working models to trace and reproduce the original work, and also use the model or data to support their own concepts, assumptions, and analyses (Cohendet and Steinmueller, 2000). Although codification does not render the tacit dimension totally irrelevant, some face-to-face interactions with human experts can be substituted by interactions with artifacts[22] (Popper, 1994; Wenger, 1998). As a result, given initial investments in developing requisite skills in the relevant scientific disciplines (the 'codebook') and the presence of some members familiar with the codebook, receivers of codified knowledge can participate in producing and developing new knowledge. Since codification externalizes knowledge, it also allows firms to acquire more knowledge than before for a given (but not necessarily lower) cost and outsource activities (Cohendet and Steinmueller, 2000). Additionally, by altering the division of labor, it entails shifting the structure of the organization and adopting a flexible or modular system of knowledge production. Consequently, in this stage, the focus shifts from 'knowledge transfer' in epistemic communities operating in stable environments to accelerating knowledge production under greater uncertainty by coordinating work in dispersed locations.

Incentives to accelerate the pace of knowledge codification and innovation[23] are strong particularly in contexts of uncertainty and rapid change.

In knowledge-intensive industries plagued by short product life cycles and facing uncertainty in new technologies, domestic firms face strong pressure to accelerate innovation and seek new opportunities to enhance survival possibilities and withstand competition from multinational corporations (MNCs). The entry of MNCs from industrialized countries as players in emerging economies is precipitated by: (1) the lure of new markets and opportunities provided by a growing middle class; (2) rising competition in home markets, compelling MNCs to seek operating flexibility (Kogut and Kulatilaka, 1994); and access to low-cost knowledge resources to help them maintain competitiveness by raising the productivity of innovation. Similarly, domestic firms in emerging economies seeking to enter global markets are motivated by the desire to develop new capabilities, acquire know-how and participate independently as knowledge producers and innovators in global markets. Consequently, strong incentives exist for codifying knowledge, which in turn, permits firms to work with collaborators across borders and also to outsource non-critical activities.[24]

Knowledge Components and Real Options Heuristics

The industrialization of knowledge production requires low-cost inputs and a larger scale of production.[25] Yet, knowledge production and innovation are fraught with uncertainty both from a technological and market perspective. One way to overcome uncertainty in the innovation process is to retain flexibility by following a *real options heuristics* strategy to limit the downside (see Surie et al., 2003 for details on real options heuristics).[26]

Since innovation requires recombining existing knowledge in new ways (Cohen and Levinthal, 1990; Nonaka, 1994), accelerating codification facilitates the creation of knowledge components necessary for the critical process of knowledge recombination (Nonaka, 1994). Various knowledge components are created at different stages of the technological life cycle (development, commercialization, adoption, erosion)[27] and can be viewed as knowledge assets (Winter, 1987). Examples of knowledge assets that can be regarded as intermediate components in the innovation process include tools, methodologies, patents, products, and services.

Moreover, since knowledge codification permits its valuation, knowledge assets created at each stage of the technological life cycle can be treated as real options (Dixit and Pindyck, 1994). This allows the firm to gain new information through experimentation and learning-by-doing while proceeding down the innovation path and deferring the decision to continue or discontinue investment until greater certainty is achieved. For example, R&D at the discovery stage of the technological life cycle that leads to the creation of knowledge assets such as tools, methodologies or patents can be licensed to other firms at interim stages to gain access to complementary resources. Likewise, the firm can establish rights to the knowledge embodied in a patent or product when entering a joint venture for development or alliance to develop a new technology (patents and R&D can be viewed as real options; Schwartz, 2003; McGrath and Nerkar, 2004). Similarly, research produced by third parties can be valued and treated as a real option. Thus, at each stage of the technological life cycle, the firm can choose to exercise the option to continue investment or terminate through exit options[28] or take advantage of new emergent options (Bowman and Hurry, 1993) such as an alliance or acquisition to co-develop or to commercialize the technology, or opportunities to license technology at later stages of the technological life cycle. Thus, in situations where there is a high degree of uncertainty, firms can gain more information about the technology and market by making small, incremental investments and regarding each intermediate asset or opportunity as a real option.[29]

New organizations lacking knowledge and complementary resources are likely to adopt a real options approach, either implicitly or explicitly. Research on new ventures in knowledge-intensive industries such as biotechnology suggests that the model followed by new firms for capability-building and growth may be characterized as one that favors 'open innovation' (Chesbrough, 2003) and sources knowledge externally rather than relying solely on internally generated knowledge. Open innovation, in turn, is possible because of the availability of tacit knowledge via licensing, since know-how is often bundled with complementary inputs in technology packages (Arora, 1996).[30] Evidence from new biotechnology firms (NBFs) indicates heavy use of alliances for expansion over the technological life cycle (Oliver, 1994; Powell, Koput, and Smith-Doerr 1996). Thus, greater availability of upstream technologies through licensing from industrialized countries enables firms in emerging economies to acquire advanced and complementary technologies more readily.[31] These collaborations also enable firms in emerging economies to build capabilities and participate in global markets as sellers of services, components, and/or final products. Examples include the provision of customized software services and components such as patents developed internally or in collaboration with other firms.

Evolving Trust in New Networks

As in the first stage when learning from suppliers required establishing trust, the adoption of a real options heuristics strategy is also facilitated by evolving trust in new cross-border networks and adopting innovations in organizational design to align with strategy.

Although codification and modular systems permit the geographical dispersion of innovation, lack of trust among partners in negotiations regarding the appropriation of knowledge can impede knowledge-sharing. Moreover, the dissemination of knowledge can reduce the competitive advantage of the supplier firm (Kogut and Zander, 1995; Gulati and Singh, 1998), particularly in 'winner-take-all' markets.[32] Consequently, the ability to appropriate knowledge is relevant. Knowledge is shared within communities when participants adopt a common code, norms, and practices signaling trustworthiness and credibility. Therefore, the transfer of tacit knowledge across borders is generally accomplished through the establishment of subsidiaries or hierarchical expansion. However, the rapid rise of cross-border innovation communities with participants from both industrialized and emerging economies coordinating knowledge-sharing across national and firm boundaries presupposes that trust exists (Arrow, 1974). Even though these communities are dispersed across borders, trust-based relationships exist between firms in industrialized countries and firms in emerging economies for three reasons.

First, a history of 'learning by doing' enabled firms in developing countries to participate in knowledge transfer and adopt the norms of multinational firms through interactions with individual experts in innovation communities of industrialized countries (Saxenian, 1994; Almeida and Kogut, 1999). In addition, past experience with technology collaborations and joint ventures has familiarized domestic firms with organizational practices of multinationals.

Second, local firms participating in cross-border knowledge production have gained increasing familiarity with relevant global standards. The widespread adoption of quality certification programs such as the ISO coupled with information availability via the Internet has also helped to diffuse ideas of empiricism, scientific rationality, and efficiency (Zbaracki, 1998; Arora and Asundi, 1999).[33] Also, by participating in global markets, emerging market firms are also familiar with requirements imposed by global institutions such as the World Trade Organization (WTO).

Third, a key factor in building a trust-based network is the likely loss of reputation and credibility if local firms do not respect contractual obligations in interactions with multinational firms. Thus, interactions between local and foreign firms are not viewed as one-shot, but as moves in a long-term, multi-stage game. Consequently, sharing knowledge and building trust in interactions can be regarded as the price of participating in cross-border innovation. Additional mechanisms for establishing and maintaining trust have also been established in these cross-border networks. An important mechanism is the use of contracts that allow participating firms to negotiate and assign ownership of the knowledge generated before engaging in collaborative knowledge production.[34] Traditional methods of protecting knowledge such as patenting are also used particularly when the knowledge shared is proprietary or considered a primary domain of activity for the participating firm.

Aligning Organizational Design with Strategy Via Heteromorphic Organization

Accommodating rising technological complexity and environmental uncertainty requires innovations in organizational design that enable coordination within increasingly complex systems. In addition, facilitating a real options heuristics strategy dictates the adoption of a matching organizational design to enable the evolution of complexity (Simon, 1956). Consequently, instead of pure hierarchies or markets, hybrid organizational solutions may be required over time to permit operating flexibility and allow the achievement of a variety of strategic goals.

By enabling codification, ICTs also permit the use of multiple organizational designs. For example, firms can adopt organizational design principles such as modularity (Baldwin and Clark, 2000) more easily. Modularity helps to encourage boundary-crossing interactions between different communities of practice,[35] promote experimentation, and enable speedier problem-solving and innovation.[36]

Modular systems are composed of units (or modules) that can be designed independently but still function as an integrated whole. Modularity is achieved by partitioning information into design rules that are visible while keeping other design parameters hidden (ibid.). Visible rules are decisions that affect subsequent design decisions and involve the architecture, interfaces, and standards. The architecture specifies the modules composing the system and their functions. Interfaces provide parameters for the interactions of modules. Standards can be developed for testing a module's conformity to the design rules and for measuring one module's performance relative to another. Hidden design parameters involve decisions that do not affect the design beyond the local module. For example, in automobile manufacturing, modular organization allows complex processes to be split up among many factories and even outsourced to different suppliers (ibid.). By analogy, modular organizational systems can also be adopted to accelerate knowledge production through distributed innovation.

However, hierarchies may still be necessary to appropriate knowledge as noted earlier, for example when commercialization is at hand. Thus, depending on the context and stage of the technological life cycle, using a real options heuristics strategy while engaging in cross-border knowledge production entails a variety of organizational mechanisms ranging from hierarchies to alliances and contracts. A consequence is the emergence of *heteromorphic* organizational form (H-form organization),[37] one that uses a combinatorial approach to structure and consists of multiple structures *over time* to match the production of knowledge components with the appropriate coordination mechanism. An organization may begin as a hierarchy and evolve to a modular structure. Conversely, an organization may adopt a more flexible structure at the beginning and then revert to hierarchy. Thus, organizational form itself is an experiment and evolves over time to match strategic goals (see Figure 2.3 for an elaboration of this model).

Combining a real options approach with the creation of knowledge assets reveals insights about the formation of markets for knowledge. It also suggests how rules evolve for linking cross-border communities of practice and generating inter-organizational networks. A consequence of the circulation of new ideas, technologies, and best practices in these networks is that local firms are able to acquire new skills, benchmark performance, adjust aspirations and adopt a new identity, that of 'world class' firms, and emerge as new players in the global economy.

CONCLUSION

In summary, the conceptual framework developed in this chapter suggests that organizational evolution occurs in stages and is dependent on the environmental and historical context at founding. In the first stage, local firms acquire technology and knowledge from external sources through learning via apprenticeship. As knowledge is institutionalized and diffused in the local environment, aspirations are adjusted and firms attempt to expand domestically and overseas. In later stages, the evolution of organizational capabilities is accelerated by the de-contextualization of knowledge production. This is accomplished by industrializing the creation of knowledge components and facilitating their exchange for complementary resources via a real options heuristics strategy and heteromorphic organizational form (H-form organization). Existing organizational structures in established firms are reconfigured to shift away from integrated modes of production that emphasize sequential information processing to more flexible modes that enhance information-processing capabilities at each stage in the evolution of the firm. While organizational inertia may delay a costly and difficult shift in established organizations, flexible structures may be more readily adopted at founding in new organizations participating in a global economy. As a result of adopting more flexible and modular design approaches to grow and expand, a new organizational form, heteromorphic organization

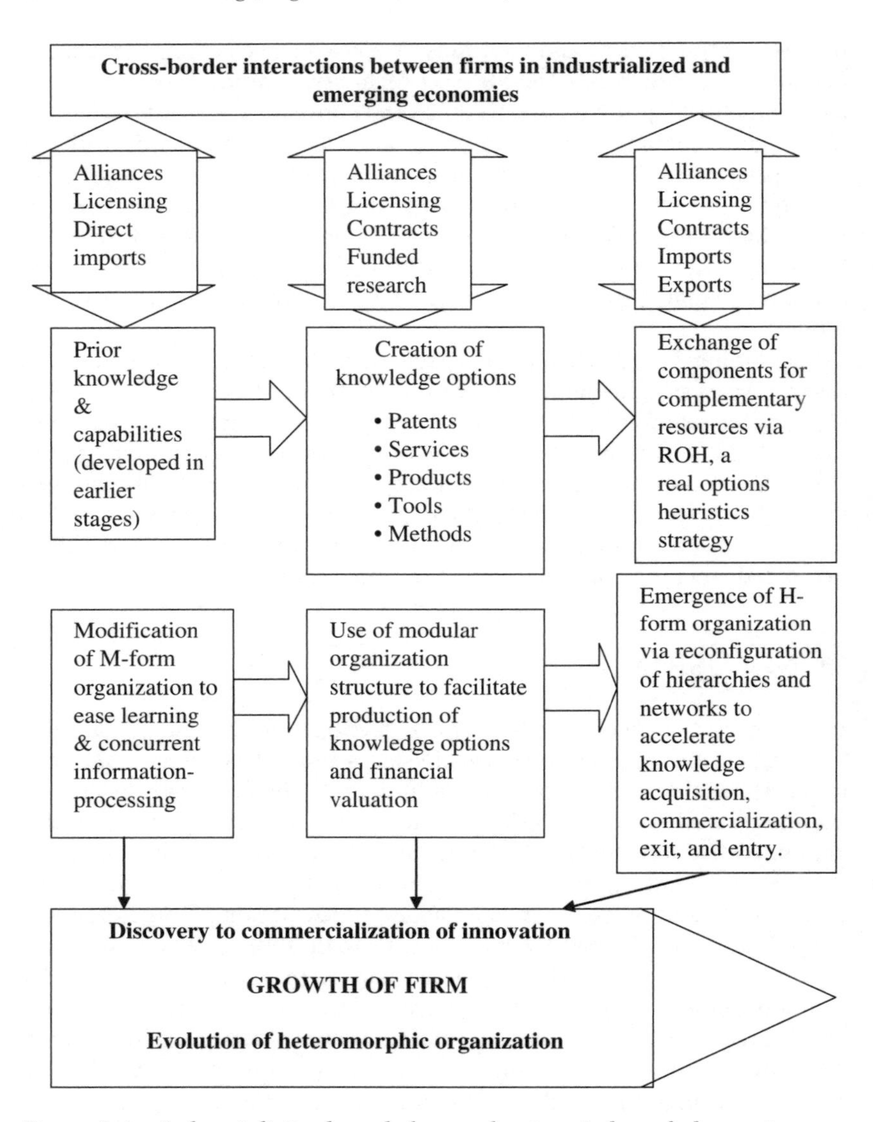

Figure 2.4 Industrializing knowledge production via knowledge options and the heteromorphic organization

(H-form organization) – including hierarchies, alliances and contracts – emerges over time, enabling the firm to maintain flexibility and gain access to new resources. It also allows firms to exploit current capabilities while simultaneously exploring the uncertain terrain of innovation.

NOTES

1. 'Knowledge' in this book refers to skill, technique, art encompassing both science and technical aspects as well as managerial and human aspects (see Hall and Johnson, 1970).
2. Hymer's ([1960] 1976) theory of internationalization as a consequence of ownership advantage is given further impetus in knowledge-based theories of internationalization (Kogut and Zander, 1993) where knowledge is the basis for ownership advantage. The present discussion focuses, instead, on how such knowledge-based ownership advantages arise.
3. Evolution is used to denote change over time; the evaluation and judgment of whether evolution is progression in a moral sense is not addressed here.
4. As noted in the introductory chapter, evolution does not necessarily imply that progress is inevitable, nor does it preclude regress.
5. As Stiglitz (2004, p. 24) notes: 'development represents a far more fundamental transformation of society, including a change in preferences and attitudes, an acceptance of change and an abandonment of many traditional ways of thinking'.
6. Reddy and Zhao (1990) provide a useful review of various strands of research on international technology transfer ranging from the impact on the home country (Mansfield and Romeo, 1980; Mansfield et al., 1983) and host country (Todaro, 1985), MNCs as possessors of quasi monopolistic advantages influencing transfer (Hymer [1960] 1976; Kindleberger, 1969), adaptation of technology by MNCs (Davies, 1977), choice of technology to transfer (Vernon, 1966), host country regulatory policies (Katz, 1985; Lall, 1985), mode of technology transfer (Hufbauer, 1966; Davidson and McFetridge, 1985) to determinants of adaptation and effective technology absorption (Dunning, 1981; Dahlman and Westphal, 1982) and transfer costs (Teece, 1977). In this research developing countries are largely viewed as recipients of technology from developed countries such as the United States.

   Although historians of science and technology (Raina and Habib, 1996) have noted the existence of indigenous technologies and argued that Indian scientists made many independent contributions to science, for the purposes of this book, I focus on technology transfer as an important method of knowledge acquisition for industrialization. Such borrowings are not unprecedented; active trade between ancient India and Western Asia suggests that the circulation, transfer, and adoption of new knowledge is normal in flourishing states (Coomaraswamy, 1965). This includes borrowings such as the Western notion of linear time (for an excellent account see Sarkar, 2002, pp. 10–37).
7. Chandler (1962) notes the importance of structure and design to accompany strategy; similarly, based on the work of Simon (1956), Cyert and March (1963) and Nelson and Winter (1982) suggest that routines are the mechanism by which structural changes are achieved. Davidson (1983) also indicates the importance of structure in international technology transfer.
8. This observation was made earlier by Cantwell (1989) but has been insufficiently explored.
9. This is akin to processes of normative and mimetic isomorphism described earlier by DiMaggio and Powell (1983).
10. See Raina and Habib (1996).
11. See Lave and Wenger (1991) and Wenger (1998) for further details of how learning is based on the creation of social communities and the characteristics of communities of practice.
12. This is akin to Nonaka's (1994) distinction between tacit and codified knowledge in his discussion of knowledge creation. Participation can be viewed as necessary for the transfer of tacit knowledge while codification accelerates its transmission.
13. Clark and Fujimoto (1991) note the importance of bi-directional feedback in new product development projects. Similarly, Daft, Lengel, and Trevino (1987) note that face-to-face or high bandwidth communication is necessary for the complex tasks of management.

14. As noted later in this chapter, power is not static but shifts with the circulation and diffusion of knowledge. Appropriation by others in the local community may cause power shifts away from experts, particularly as the organization expands.
15. While Weber's (1946) seminal insights on power and authority remain relevant, the context of learning requires a corresponding shift of power to the learner as he or she gains mastery.
16. The craft mode of production refers to an apprenticeship model as described in Piore and Sabel (1984) and Lave and Wenger (1991).
17. Characteristics of such communities identified by Wenger (1998) include: (1) sustained mutual relations, (2) shared ways of engaging, (3) rapid flow of information and propagation of innovation, (4) absence of introductory preambles, (5) quick set up of problems to be discussed, (6) substantial overlap in participants' descriptions of who belongs, (7) knowing what others know, what they can do and how they can contribute to an enterprise, (8) mutually defining identities, (9) specific tools, representations and artifacts, (10) shared stories, jargon, and shortcuts to communication.
18. Popper (1994) notes that the growth of knowledge involves testing assumptions and theories; the result is the emergence of new assumptions and theories.
19. This process is similar to Nonaka's (1994) knowledge creation cycle in which tacit knowledge is first articulated and codified, and, subsequently re-embedded in routines so that it becomes tacit. Well-known examples include studies by Bartlett and Ghoshal (1988) and Ghoshal and Bartlett (1990) on internal transfers within multinationals like Philips and Matsushita; an early version of organizational practices required for lean production are described by Aoki (1990) in his work on the Japanese form of organization.
20. 'Codebook' refers to the specialized language of the community. For example, each domain of science has its own vocabulary and terms that only members of the community understand (Cowan, David, and Foray, 2000).
21. A modular organizational structure (Baldwin and Clark, 2000) is one that permits collaboration by many individuals, yet permits each participant to work individually.
22. These include documents, working models, methodologies, tools, concepts, and so on.
23. Since innovation depends on recombining knowledge, it is linked with the pace of knowledge codification (Nonaka, 1994).
24. The trend of outsourcing activities is prominent and continues to accelerate in industrialized countries.
25. Chandler (1990) attributes the rise of the modern corporation to the industrialization of production, which necessitates the use of technology to increase the scale of operations and accelerate throughput. By analogy I apply these ideas to knowledge production.
26. A real options heuristic is a guide for making strategic investment decisions under uncertainty by treating such investments as real options. Real options are investments resembling financial options whose value is a function of volatility and time to expiration (Kogut and Kulatilaka, 2001). As volatility or general uncertainty associated with an investment increases, the investment is discounted because of risk aversion, while the reverse is the case for a call option that increases in value as volatility increases. Since the downside risk is limited to the cost of the option, increased volatility increases the chance that the price of the underlying asset may exceed the exercise price before expiration without increasing the downside risk (Mitchell and Hamilton, 1988).
27. These stages are well recognized in research on technological innovation and its diffusion (Rogers [1962] 1995).
28. Schwartz (2003) notes that the abandonment option represents a large portion of the project's value when the project is marginal and/or when uncertainty is large. A termination strategy often involves using joint ventures. Kogut (1988; 1989) and Bleeke and Ernst (1991) note that the majority of joint ventures are terminated through acquisitions.
29. While problems of valuation are well known, I focus on the decision-making and information-gathering aspects of using a real options heuristics strategy, particularly under uncertainty. See Surie, McGrath, and MacMillan (2003) for an elaboration of real options heuristics strategy.

30. Using data on 139 leading chemical technologies, Arora, Fosfuri, and Gambardella (2001) found that the presence of technology suppliers operating in industrialized countries increased investments in chemical plants in less developed countries. They argue that upstream specialized technology suppliers can benefit downstream firms by licensing technologies, especially those that are technologically less advanced.
31. This is corroborated by Acs, Audretsch, and Feldman's (1994) finding in the United States that small firms' innovative capacities depend on investments that other firms and institutions make in R&D. Thus, small firms benefit from R&D spillovers.
32. See Noe and Parker (2005) and Salganik, Sheridan Dodds, and Watts (2006).
33. This is not to imply that scientific ideas of empiricism and rationality did not exist previously. However, the wider dissemination of these ideas is accomplished through the adoption of ISO-type standardization.
34. Stiglitz (1989; 2004) noted that an impediment to development is the paucity of appropriate institutions to support markets. The transformation from 'status' to 'contract' in developing economies is profound (Lewis, 1970).
35. Communities of practice include such groups as those involved in buyer–supplier relationships and interactions across borders and within each country. Communities of practice also span relationships and interactions between different groups within the firm such as product design and marketing.
36. Modularity offers advantages in design because experimentation and improvement of components can occur within a module, behind an interface, without compromising the efficacy of other components and without requiring a change in overall system architecture. Likewise, architectural changes can be accommodated with minimal effect on individual components. This is advantageous when the complexity of the product requires the collaboration of many, none of whom can know all of the design individually. Hence, responsibility for component, interface, and system architecture design can be distributed. Local decisions about components need not affect the overall architecture and system. Similarly, project managers can alter system architecture without reducing individual component functionality. In the absence of modularity, component experimentation must be more tightly integrated and coordinated because of interactions between modules and the system-wide effect of these interactions (Baldwin and Clark, 2000). Thus, modularity provides a way to divide overall system labor and knowledge so that individuals responsible for the system, components, and sub-systems can experiment and make improvements independently.
37. The word heteromorphic refers to an evolving organism, one that changes its state and encompasses two or more states simultaneously. Heteromorphic (H-form) organization emphasizes flexibility and blurs the boundaries between markets and hierarchies; in contrast multi-divisional (M-form) organization (Williamson, 1975) emphasizes hierarchy as a means of gaining efficiency via minimizing transaction costs.

3. Methods

This book uses a case-study approach to examine how domestic firms in an emerging economy evolve through interactions with foreign firms. The data collection and analysis were conducted in stages from 1993–2007. In the first stage, the study covered ten firms in the manufacturing sector, including construction equipment, steel, automotive components such as bearings and pistons and rings, power equipment, motorcycles, agricultural equipment, and engines for tractors, utility vehicles, and automobiles. The manufacturing sector was selected for its critical importance to developing countries, capital intensity, and relevance in international markets. In the second stage, the focus shifted to newer, high-technology industries and included firms in the software and biopharmaceutical industries. These industries were included because of the increasing importance of ICT-based industries in industrialized countries, the rising presence of Indian players in this arena and to determine whether these firms followed a similar evolutionary path to manufacturing firms. Firms from both manufacturing and high-technology sectors were revisited in the third phase to obtain a broader perspective of the economy.

A case-study approach was deemed relevant because of the nature of the research questions. Initially, the objective was to focus on organizational processes to explain how learning occurred in technology transfer situations and why some firms absorbed technology more readily than others. In later phases of the study the objective was to understand how expansion and internationalization occurred and whether firms in different industries employed similar processes and followed similar evolutionary paths. The Indian environment provided a real-life context in which to examine a 'natural experiment' that was most readily accessible via a case-study strategy. Multiple sources of evidence were included, ranging from direct observation to systematic interviewing to obtaining data from public and private archives. Moreover, since context is important, any fact relevant to the stream of events describing the phenomenon is a potential datum in a case study (Yin [1989] 1994).

In studying the evolution of these firms it was necessary to examine the phenomenon of learning and growth from varied perspectives. Individuals working at different levels and functional areas within these organizations were interviewed to understand interacting factors: the engineers, managers, and executives involved in introducing and implementing new technology projects, formulating and implementing strategy, conducting R&D, and bringing new products to market. Consequently, the use of a case-study approach was considered appropriate to facilitate '*Verstehen*' (understanding).

RESEARCH DESIGN AND CASE SELECTION

A multiple-case design was employed to allow for comparison across cases and ensure that the emerging theory could be extended beyond insights offered by a single case. Multiple cases also help to guard against potential biases such as misjudging the representativeness of a single event (Tversky and Kahneman, 1986), increasing the salience of a datum because of availability, or biasing estimates because of unconscious anchoring (Leonard-Barton, 1990). Cases were selected to permit literal replication (cases for which similar results are predicted), or theoretical replication (cases for which theory predicts contrary results), an approach that is similar to that underlying multiple experiments (Yin [1989] 1994, pp. 52–4) and one that would permit the development of a theory relevant to different situations and capable of explaining both success and failure.

Exploratory interviewing in the early stages of the study in 1993 revealed three potential sites (at firms that are part of one of the largest conglomerates in India) where foreign technology had been adopted. The first case, a construction equipment firm, had transferred technology successfully (via technical collaboration or licensing in which local management retained control over technology implementation). As one of the oldest engineering firms in India with a long history of technical collaborations, it was selected as an ideal case. The second case, the bearings firm, was included to allow for theoretical replication since it was a transfer by joint venture. It was anticipated that the proprietary nature of the technology would inhibit transfer. Literal replication was also sought in the third case, a steel firm, in which transfer was accomplished via direct import of technology and equipment. This firm was selected because it was one of the oldest firms in India and with a history of technology transfers was likely to have developed routines for technology transfer. Moreover, it was expected that using the appropriate organizational processes would be as relevant in a continuous process manufacturing firm as in batch manufacturing firms. The following criteria were used to select the three initial cases:

1. The recipient firms were large, established companies in India.

2. The industries were similar (occupations and key operations employed were similar – machinists, drafters, machine tool cutting operators, lathe and turning machine tool operators; key operations included grinding, heat treatment, turning. (Industry – Occupation Matrix, US Bureau of Labor Statistics, in Darnay, 1993).
3. Project sizes were similar (over INR50 million).
4. All the firms employed mature technologies.
5. Transfers were to the same location, allowing for control of regional effects since all three firms draw on the same labor pool and hence have access to similar levels of technical skill.

The key difference between the three cases was the mode of transfer. Each firm used a different mode: technical collaboration, direct import of equipment, and joint venture.

The same logic was followed in selecting the seven remaining cases from the manufacturing sector. They were included because the technology transfer projects were accomplished either by joint venture, technical collaboration, or direct import during the same time period (1980s onwards). It was expected that when conditions matched those outlined in the framework (Stage I; organizational processes and systems match task demands) transfer is more easily accomplished, thus providing literal replication. Alternatively, when different conditions prevail, a match between processes and task requirements is not achieved and knowledge is not easily shared, transfer is more difficult, leading to theoretical replication. A two-step process was followed to identify these firms.

First, firms in the manufacturing sector that were potential recipients of technology were identified using three sources: (1) a directory published by the Confederation of Indian Industry (CII; 1994) listing members and company information; (2) a directory listing foreign collaborations in different industries, which was used mainly to identify technology recipients; (3) a list of likely candidates in the Delhi region provided by the Industrial Credit and Investment Corporation of India (ICICI). Second, based on this information, a shortlist was developed and, with the help of the CII, letters were sent to the managing directors or senior executives of each of the targeted firms requesting information about the transfer and permission to conduct the research. The target firms were among the top Indian companies in terms of sales turnover and in most cases, the suppliers of technology to these companies ranked in the top ten in their respective industries. Of the 11 companies initially identified as suitable, five firms in and around Delhi were finally included in the study, one firm in the Western region and one in the South in addition to the first three cases in the Eastern region. The rationale for varying the geographical location was that it would be possible to examine whether the emerging theory was robust in different contexts. Finally, depth of access, of critical importance in field research, was an important consideration in site selection, particularly in the three initial cases.

In 2000, I conducted further interviews at the three firms initially selected for the study. In a later phase of the study (2003–07) the objective was to examine the phenomenon of international expansion via cross-border innovation. To do so, it became necessary to study innovation-intensive firms and their interactions with multinational firms. Again, a multiple-case design was employed to examine two high-technology industries – biotechnology and software – to facilitate cross-case comparisons (Glaser and Strauss, 1967; Strauss and Corbin, 1990). Both industries are knowledge- and innovation-intensive, and thus provide an appropriate context for examining the research question (Pisano, 2000). Additionally, changes in the Indian environment yielded a setting in which to examine the effects of uncertainty. Both industries are important globally. The United States is a leader in both industrial sectors, which are strong contributors to cross-border trade (biotechnology trade grew by 13.2 percent on average from 1996–99), important recipients of venture capital, and active in patenting (in the United States, software-related patents account for 4–10 percent of all patents). Both industries are also gaining importance in India. The Indian IT industry, which began in the 1970s, is expected to grow to US$77 billion by 2008 (NASSCOM-McKinsey Report, 2002). While biotechnology is an emerging industry in India, it is expected to attain a global market share of 10 percent in the next five years and is being actively promoted by the Indian government.

As before, online lists of biotechnology and software companies from two states in India, Andhra Pradesh and Karnataka (where the majority of the leading biotechnology firms are located) were developed in 2002 and firms were selected using theoretical sampling (Eisenhardt, 1989; Yin [1989] 1994). Firms were contacted by e-mail and a total of 18 organizations were selected. These included six software firms, eight biotechnology firms, two associations (one in each industry), a life sciences laboratory in a university, and the industry ministry of a state promoting both industries. Two of these firms were software subsidiaries of US multinational corporations (MNCs) and two were R&D centers of global pharmaceutical firms. The rest were Indian firms. It was anticipated that by including a diverse array of industry participants and network actors it would be possible to obtain a variety of perspectives and gain a fuller picture of each industry. In 2007, several of the firms that were part of the previous two studies were contacted. Among these were four manufacturing companies, a biotechnology firm and a biopharmaceutical company. In addition, some new ones were opportunistically included. These included a heavy commercial vehicles (trucks and buses) manufacturer and mining services firm in Southern India. Moreover, selected scientific institutions, industry associations, and government departments were also included to gain an understanding of the state of science and technology and R&D capabilities in India. The names of all organizations have been disguised for confidentiality.

DATA COLLECTION

Data were obtained from multiple sources including semi-structured interviews, documents such as annual reports, company newsletters, press releases, internal reports, published sources, and websites. A survey questionnaire was administered for the initial study. The paucity of responses despite tremendous follow-up led to abandonment of this strategy in favor of interviews. Interviews were semi-structured, conducted at the research site, and used a protocol following guidelines outlined by Eisenhardt (1989) and typically ranged from one to two hours. Interviewees were CEOs, senior executives, or R&D heads in both Indian firms and MNCs.

I conducted about 85 formal interviews from 1993–2000. During this time, I also attended meetings at ten sites, conducted workshops, and taught executives in two leading business schools. This immersion in the environment gave me a rich understanding of the changing context. In 2003, I conducted approximately 30 interviews followed by another 30 interviews in 2007. Direct observation of researchers at the various sites, presentations, and facilities tours with members of R&D organizations helped provide a 'real time' aspect to the research and first-hand understanding of the kind of technology used and how research was conducted. Following strict case-study protocol, with the exception of about 12 interviews, all interviews were audiotaped and transcribed. Some data were also collected via questionnaires.[1] Interview notes were taken in most instances and were used when tapes were unavailable. The interview data were triangulated by cross-checking information against archival and public documents and information derived from informal meetings. In addition, while establishing rapport with interviewees, I took note of my own responses to interviewees as well as the location and context of the interviews to remain aware of interviewer bias.

Data collection was continued to the point of 'saturation of categories' and the emergence of regularities (Miles and Huberman [1984] 1994). While these methods permitted me to develop a keen sense of the context and organizational processes, it cannot be said that systematic random sampling was pursued.

DATA ANALYSIS

Since statistical methods were inappropriate, a more qualitative approach was used following guidelines recommended by Miles and Huberman ([1984] 1994), Yin ([1989] 1994) and Strauss and Corbin (1990). Analyzing case-study evidence consists of 'examining, categorizing, tabulating, or otherwise recombining the evidence to address the initial propositions of the study' Yin ([1989] 1994, p. 105). Miles and Huberman ([1984] 1994, p. 21) suggest that 'analysis consists of three concurrent flows of activity: data reduction, data display, and conclusion drawing or verification'.

Data were analyzed by reducing transcripts to categories and empirical findings (Glaser and Strauss, 1967; Yin [1989] 1994; Strauss and Corbin, 1990). The data analysis process iterated between theory and data; while initial constructs were drawn from the literature, they were not imposed. Rather, empirically based patterns (Yin [1989] 1994) were compared with past literature using a pattern-matching logic. For example, in the first study, I approached the firms with the technology transfer and internationalization literature in mind. However, as data collection progressed, the emphasis on improving quality, becoming 'world class' and 'a learning organization', led me to incorporate the literature on learning via building a community of practice to categorize findings.

In the second study on knowledge-intensive high-technology firms, the empirical data suggested that interactions between Indian firms and MNCs stemmed from the need to pursue opportunities in a new market while reducing costs and risk. The ubiquity of such partnerships suggested that each firm[2] was pursuing a strategy to maximize options and led me to incorporate the real options literature. Involvement in another study on biotechnology and software firms in the United States (findings not reported here) led me to confirm the relevance of real options for this framework. Thus, following the methods of inductive theory development, I next focused on mechanisms that enabled firms to follow a real options heuristics strategy and examined situations that either fit or did not fit these categorizations. Based on analyses of focal firms in the biotechnology and software industries that were following an expansion strategy, I then used analytical replication to determine whether the framework developed was confirmed or disconfirmed in the rest of the organizations. I developed tabular arrays to draw comparisons across cases (Eisenhardt, 1989 and Miles and Huberman [1984] 1994). These analyses helped to construct a framework based on knowledge for internationalization and the evolution of firms and markets.

In the last phase of the study conducted in January 2007, the data confirmed the expansion and evolution of domestic firms. After revisiting some of the manufacturing firms as well as some of the high-technology firms, it appeared that firms in both industries were pursuing a similar approach and faced similar challenges. The emphasis on growth and the need to achieve scale were common themes in most interviews. In addition, a common feature was the use of standardization in achieving scale while trying to retain flexibility, leading me to focus on how complex systems evolve and adapt. Hence, I incorporated the literature on complex adaptive systems into the framework. Consequently, the overarching theoretical framework presented in the last chapter is a synthesis of various theoretical approaches that emphasizes the primacy of aspirations, experimentation, and entrepreneurship.

The patterns deciphered are presented through the prism of focal cases to preserve the richness of context while illustrating the framework.

LIMITATIONS

While the case-study design and qualitative methods were deemed appropriate for studying cross-border learning and evolution of firms, there are some limitations. The vulnerability of the data to subjective interpretation and difficulties in analyzing voluminous amount of data are well known (Leonard-Barton, 1990; Miles and Huberman [1984] 1994). The time-consuming nature of multiple case-studies and labor intensity of the data collection process (given that it requires establishing rapport and developing relationships at various research sites) is evident. In addition, the research must be coordinated to ensure that insights derived from data collected by one particular method are used for further investigation with another method. Although using more than one researcher can alleviate subjectivity, a strong rationale for relying on one researcher is that experience yields deeper insights that are more coordinated and synthesized. Hence, while multiple researchers were not feasible for data collection, the ideas were tested and reviewed at various stages by experts and peers in academia via presentations at conferences both in the United States and India (Miles and Huberman [1984] 1994 suggest that reviews help to ameliorate some of the problems in interpreting data). Another limitation is that it may be well into the data collection phase that patterns begin to emerge and major issues become apparent. Thus, issues that emerge in subsequent cases can only be addressed by further in-depth investigation.

Finally, there must be a match between researcher skills and methodology selected. Interviewing skills, a high tolerance for ambiguity and sustained efforts are required; moreover, time and effort are required to set organizational expectations and foster relationships to ensure cooperation. Another researcher may have accomplished these differently.

Nevertheless, the approach adopted here was appropriate because of the nature of the study, focus on organizational processes, and permitted understanding of the context. Since the purpose was to develop a theoretical framework from patterns that were grounded in data, hypotheses derived from such research must be tested through large sample studies.

NOTES

1. Interviews were not recorded for the following reasons: (1) when permission was not granted, (2) operator error in handling the recorder, or (3) when interviews occurred unexpectedly and a recorder was not available. Additionally, some interview tapes were stolen along with the micro-cassette recorder. Transcriptions for the first study were done by the author and a student trained by the author. For the later studies, interviews were transcribed by a student and by a transcription service in India. Data were collected via questionnaires for the following categories: coordination, integration, associated knowledge of engineers and reliance on foreign expertise (see Surie, 1996 for details). The questionnaire was used in the three focal firms to supplement and corroborate interview data. However, it was abandoned because of the difficulty of obtaining mailed responses from other firms.
2. This applies to both Indian and multinational firms.

4. Knowledge transfer via apprenticeship in Indian manufacturing firms: Stages I and II

This chapter presents evidence on the first two stages of globalization from three longitudinal case studies of technology transfer projects. These studies were replicated in seven other firms. The first phase of the study from 1993–96 focused on knowledge transfer. The second phase (1999–2000) of the study focused on institutionalizing learning.

The first case, Earthmovers, a construction equipment firm, was selected because of its previous collaboration with a US supplier to manufacture mechanical excavators in the 1960s and 1970s. In the mid-1980s, Earthmovers entered a technical collaboration with a Japanese supplier to upgrade its technology and produce hydraulic excavators to compete with domestic and international manufacturers. Transfer through arms' length collaboration suggested that establishing a community of practice could be difficult, despite prior experience with technology transfer.

The second case, Bearings, Inc., provides an example of transfer by joint venture between a leading Indian steel company and a premier US bearings manufacturer. Difficulties in establishing a community of practice were anticipated for two reasons: (1) the foreign partner's desire to protect proprietary technology was expected to inhibit practice and knowledge-sharing; (2) the US partner's lack of prior experience in the new context was likely to lead to the replication of established routines with little attempt to adapt to the new context (Kogut and Zander, 1996).

The third firm, Steelworks, the oldest steel company in India, established in 1907, was also selected for its experience in technology transfer. However, Steelworks chose to import technology directly. Instead of developing a long-term relationship with a single collaborator, Steelworks drew on a network of suppliers, consultants, and industry specialists to install an automated state-of-the-art blast furnace purchased from Portugal. Consequently, I expected a local community of practice to evolve, enabling some learning, but not necessarily one aligned to global firms.

Evidence from each of these cases is presented longitudinally using the framework outlined in Chapter 2 to highlight the evolution of capabilities and knowledge stemming from technology infusion from industrialized countries. In Stage I, the craft mode of knowledge transfer and replication necessitates the creation of epistemic communities of practice. This involves:

1. fostering relationships;
2. organizational design for knowledge transfer (a shift to concurrent organizational design to promote problem-solving);
3. creating opportunities for learners to practice skills in context;
4. mastery and evidence of knowledge transfer.

Case presentations are organized according to these categories. The successful creation of an epistemic community results in the replication of knowledge and evolution of capabilities in the new context. Stage II focuses on institutionalizing learning and replicating it across the firm. These findings were corroborated by evidence from seven other manufacturing firm presented using the same categories in the last section of this chapter. (See Table 4.1 for a summary of knowledge transfer processes in Stages I and II for the first three cases and Table 4.2 for the seven other firms.)

Stage III is discussed in the next chapter. As these firms introduce new technologies and continue to reorganize and modularize organizational structures based on past technology infusions, it becomes possible to shift from adaptation to innovation and autonomous knowledge creation.

STAGE I: TECHNOLOGY TRANSFER

Technology Transfer in a Construction Equipment Firm: Apprenticeship in Action

Earthmovers Limited, a leading engineering company in Eastern India faced a major transformational challenge – going from being a small domestic manufacturer of excavators and cranes to a globally competitive manufacturer of construction equipment. Such a transformation required upgrading skills and capabilities, not just expanding capacity. A pioneer in the engineering industry and promoter of industrialization (Earthmovers was established in the 1940s and is part of a leading national conglomerate), the company had maintained its leading position by operating in a framework emphasizing indigenization, import substitution, infrastructure building, and social welfare in keeping with its identity as a nation-building firm.

The liberalization of the Indian economy in 1991–92 resulted in competition from global companies. Hence, acquisition of capabilities and technological mastery became critical for survival. Earthmovers' history of technological collaborations with firms in the United States and Germany had prepared it for such changes to come. Established in the 1950s as a separate unit, the excavator division initially manufactured spare parts for cranes and excavators imported from the United States to use the spare capacity of the steel foundry and general engineering division. A decade later it signed a collaboration agreement with a US supplier to manufacture two excavator models in India. Initially imported and assembled from imported CKD[1] kits, by 1961 they were manufactured locally. Output rose from three machines in the first year to 33 by the end of the second. Additional models were introduced over the next ten years and the technology agreement was extended. Concurrently, local facilities and capabilities were upgraded to include material processing, fabrication, machine shops, and heat treatment. By 1975, marketing and service functions were brought under direct control to gain access to feedback from customers, prevent delays in communication, and permit pricing flexibility.

Faced with competition from domestic rivals who started manufacturing hydraulic machines, some in collaboration with leading global firms, and realizing their existing technology was obsolete, Earthmovers decided to discontinue making mechanical machines and manufacture hydraulic machines. To speed market entry, indigenous design and manufacture – the option initially pursued – was abandoned in favor of collaboration with a Japanese firm in mid-1984 to purchase technology and use the Japanese brand name jointly with its own name to market the new products in India and internationally. The new hydraulic excavator was introduced within a year and three additional models were introduced by 1985–86, while some older products like mechanical shovels were phased out and others like cranes were upgraded. Over the next five to seven years, Earthmovers consolidated its position domestically and regained its leading edge by increasing output via mechanization, reducing imports and reducing costs by manufacturing and purchasing components locally.

Dependence on imports was reduced through indigenization and stabilization of production processes. In 1985, the model with the largest volume had an import content of 40.9 percent (INR412 000 CIF[2] value @ INR100 = ¥2000). By 1990–91, through planned indigenization, the import content had fallen to 11 percent (INR288 000 CIF value @ INR100 = ¥850). Indigenized components included hydraulic cylinders, hoses, track links, radiators, cabs, gear boxes, and high and low pressure pipes. However, high-precision products such as hydro-motors, pumps, and control valves requiring capital-intensive equipment for manufacture were still imported.

Indigenization also helped to reduce component costs during a period of rapid appreciation of the Yen (over 135 percent during 1985–90).

Investment in capital equipment (approximately INR40 million) facilitated mechanization. This included the installation of equipment such as a multi-torch oxy-cutting machine, a CNC (computer numerical control) machining center, and turning and gear hobbing machines. Efforts focused on developing reliable sources of high-quality components.

To meet increasing customer demand for product variety and high quality at acceptable prices, two indigenously designed products, a wheel-loader and an articulated crane were introduced. Simultaneously, a new series of excavators was also introduced via Japanese collaboration to replace earlier models. Organizational changes mirrored technological changes. The 'Excavator Division' was designated the 'Construction Equipment Division' in early 1992 to signal the intent to expand as an independent entity. Another reorganization in October 1994 led to its designation as the Construction Equipment Business Unit, an independent unit with its own representative at the corporate office in Bombay, endowing it with greater autonomy and accountability.

Creating a community of practice

Four elements of community creation are examined to highlight the dynamics of knowledge transfer:

1. fostering relationships to speed communication and knowledge dissemination;
2. organizing work to promote coordination and problem-solving;
3. facilitating practice to build capabilities through apprenticeship;
4. mastery and evidence of knowledge transfer.

Fostering relationships From the perspective of participating in communities of practice, learning requires involvement and 'concerns the whole person acting in the world' (Lave and Wenger, 1991, pp. 49–50). Thus, activities, tasks, functions, and knowledge are part of 'a broader system of relations in which they have meaning' (ibid., p. 53). Learning involves constructing a new identity through interactions within the systems of relations in the community. Communicating effectively with peers is important for learning to be effective and the learner's relations to peers and experts helps to establish opportunities to learn. Hence, building trust-based relationships between participants in the transfer project to promote communication is essential for disseminating know-how relevant to practice. I broaden the concept of relationship to include actors in the network rather than focus solely on the dyadic relationship between the experts and learners.

I focus on four types of relations:

1. relations between the technology supplier and recipient;
2. inter-departmental relations;
3. relations between management and workers;
4. relations with domestic suppliers.

Managers in Earthmovers reported that their relationship with the Japanese technology supplier was open and supportive. Despite a reputation for secretiveness, the supplier responded quickly to requests for information. Japanese experts acted as mentors and helped workers to improve performance rather than censuring them for substandard performance.

Inter-departmental relations were perceived as cordial in Earthmovers, although marketing and purchasing reported poor relationships (questions for assessing relationships were adapted from Lawrence and Lorsch, 1969). Executives noted that the mechanism of dialogue had been introduced to promote information flow between internal 'customers' and 'suppliers', to increase responsibility and autonomy at lower levels of the hierarchy, and to enable engineers to 'multiply their capacity for thinking'. The head of production pointed out the need for reinterpreting organizational culture to ensure that employees understood the importance of their work:

> We are trying to induct a culture [which provides] feedback to the doer directly, which is best. But it does not always happen. We want to make it as direct as possible . . . on schedule, in a daily formalized manner . . . so that the [internal] customer gives feedback to the [internal] supplier.

Cross-functional teams and groups at the worker level also helped bypass the multi-tier hierarchical system and improve communication through face-to-face interactions rather than through written memos.[3] Managers in Earthmovers also emphasized teaching subordinates how to improve working relations with other departments and paid attention to providing subordinates with information about industry-related events and expertise.

An ancillary development department was established to train local suppliers to upgrade the quality of outsourced components.

Organizational design for knowledge transfer Problem-solving is expedited via organizational design for concurrently processing work to surface early discovery of potential bottlenecks later in the project. However, concurrent processing requires greater communication and coordination across functional units. The nature of work-flow (amount of work processed independently, sequentially, reciprocally, or via team coordination) was assessed through questions about how the design tasks were organized and coordinated.[4]

Most engineers in Earthmovers indicated that tasks were concurrently organized; the divisional manager for welding and fabrications contrasted the sequential nature of task organization required for the earlier mechanical excavators with the new organization required for building machines with a modular design and fewer parts: 'The old cranes were built brick by brick. Unless you make one part, you can't make another . . . Here the components can be made independently; the design is modular and the weight [the steel requirement] is much less'.

Increased communication dictated installing computerized systems and integration across the marketing, design, and production departments. According to the project planning manager:

> We are involved in CAPP and CAM[5] from A to Z. We help the designer to correct his designs so that they are compatible to manufacturing; then when he gives every product for casting . . . We do the material sourcing, we communicate their drawings to the forge, the foundry and to the vendors for development, so this is known as concurrent engineering.

Questionnaire data indicated that integration across functions and concurrence in project implementation in Earthmovers was 20 percent higher than that observed in Bearings and 10 percent higher than in Steelworks. Nevertheless, complete transparency was constrained by existing equipment and manufacturing capacity.

Opportunities to practice Involvement in the work of the community of practice was promoted in Earthmovers via hands-on training in India and Japan, through participation in different activities, and by adapting technology to suit local conditions. Eighteen welders were trained in Japan for one year while Japanese experts provided assistance in jigs and fixtures, plant layout, and production scheduling. Engineers visiting Japan were allowed to take photographs as visual aids for understanding new technology.

All shop floor operators were taught basic skills in machine operation, maintenance, self-inspection, and troubleshooting for minor problems resulting in reduced reliance on technical experts. Also, by adapting new techniques such as the CO_2 welding process, participants acquired *know-why* and understood the principles underlying the technology.

Mastery: evidence of knowledge transfer Mastery of objective knowledge is demonstrated by skilled performance and effective problem-solving in manufacturing, production and design, leading to innovation, new products, and exports (Westphal et al.,1985; Lall, 1987; and Scott-Kemmis and Bell, 1988). In addition, subjective knowledge transfer involves successful socialization into new perspectives (Wenger, 1998). In these transfers, adopting the suppliers' perspective involved learning new performance norms for productivity, quality, and innovation by developing a problem-solving orientation, and a quality and customer orientation, thus aligning membership and identity with global manufacturing firms.

In Earthmovers, process improvements led to a 50 percent reduction in cycle times for heat treatment operations. The autonomous design of new products such as wheel-loaders, backhoe loaders, cranes, and mini-excavators indicates that design capabilities were enhanced.

Interactions with Japanese experts resulted in adopting a problem-solving lens. The notion of diagnosis, testing, and evaluation was applied to manufacturing and dominated every sphere of activity including planning, prototype construction, quality, materials, and supplier development. Emphasis on problem-solving led to increased reliance on quantitative data and a need for systematic data-gathering and training in problem-solving techniques.

Similarly, a quality and customer orientation was adopted in tracking machine performance by region and customer to determine whether poor performance or machine failures were region-dependent or widespread. High failure rates in the case of machines supplied to granite producers led to redesigning the bucket to improve the performance of machines subjected to high levels of stress in granite quarrying. Hence, quality assurance encompassed changes in design and was also demonstrated by the achievement of quality certification (such as ISO[6] certification) in production and design capability, signaling intent to enter international markets. Subjective knowledge also evolved as organization members became conscious of their newly emerging identity as a quality manufacturer aspiring to be world class.

Exports commenced to markets like Bangladesh, Nepal, the Middle East, and Africa but competing in industrialized markets was constrained by low production volumes and the lack of development of related engineering industries, suggesting that the domestic economic environment must also be considered.

Technology Transfer in a Bearings Manufacturing Firm: Apprenticeship Inhibited

Unlike Earthmovers, which used technical collaboration or licensing as the primary technology transfer mode, Bearings, Inc., established in 1987, was a joint venture between one of the largest Indian steel companies and a premier US bearings manufacturer of tapered roller bearings and specialty steels. However, concern for appropriable knowledge by the partner supplying technology, common in joint ventures, was expected to inhibit knowledge-sharing.

The joint venture was formed to manufacture and market tapered roller bearings based on the US partner's proprietary technology.

The US partner

With sales turnover of US$1708.7 million in 1993, the US firm employed 18 000 people worldwide and operated 22 production units in eight countries – the United States, Australia, Brazil, Canada, France, Great Britain, South Africa and India. It designed bearings for original equipment manufacturers and was increasingly concerned with meeting the needs of global markets. The firm prided itself on maintaining high-quality standards that ensured complete interchangeability of bearings and bearing parts regardless of manufacturing source or country of origin. The company was a key supplier to leading global manufacturers of vehicles and industrial machinery. In addition, it was also committed to the highly competitive after-sales market for tapered roller bearings and worked closely with a global network of authorized distributors to supply replacement needs of vehicle and industrial equipment operators. It also produced high-quality alloy steels for the bearing, automotive, and aerospace industries. Like the Indian partner, concern with adapting to a rapidly changing environment led the US firm to launch an aggressive initiative to reduce costs, launch improvements in facilities and restructure operations in 1993. Research, conducted mainly in the United States and United Kingdom, focused on improving bearing performance to meet the challenges of a changing technological environment. The marketing director for Europe noted:

> Times have changed. At one time, the US was the world's economic wellspring, producing a river of new products. Today, the springs that create the river are everywhere. Our challenge is to find those sources wherever they may be and to impress the designers who are creating a new generation of products for specialized, highly segmented markets.

The Indian partner

The Indian partner, a steel company, was established in 1907 in Eastern India and is a member of a leading Indian conglomerate. The firm was conceived by a visionary who sought to industrialize India by setting up the first steel and electric power plants and the first science and technology institute. The firm was organized as an integrated steel plant with raw materials obtained from its own mines and collieries in two Eastern states. At the behest of the founder, the township was located on the banks of a river to ensure adequate water supply and designed to provide a pleasant living environment and avoid the levels of pollution found in British and US centers like Sheffield and Pittsburgh (Lala, 1981).

The Indian partner had a turnover of about INR58.4 billion (approximately US$1.7 billion)[7] in 1995–96, production of 2.7 million tonnes of saleable steel and a total workforce of 70 709 employees. The steel firm produced billets, blooms, slabs, wires, rods, bars, sheets, structurals, hot and cold rolled strips, railway tyres and axles, agricultural implements, steel plant heavy engineering equipment, and chemicals such as sulphate of ammonia and crude naphthalene, a by-product of steelmaking. Its three subsidiaries included a refractory plant, a pigment manufacturing company, and an investment company that held controlling interests in a specialty steel manufacturer of steel billets and wires used in the construction, textile, automobile, and bearing industries. The Indian partner also embarked on a diversification program into companies in three areas: steel, engineering and allied products (this included the acquisition of a bearings company), and refractories. A new cement plant with a capacity of 1.7 million tonnes was the latest diversification venture. Diversification was undertaken to enable the firm to compete more effectively in domestic and international markets in a liberalized and more competitive environment. Moreover, in every area, management was taking steps to make the firm more productive through modernization, technology upgrades, and diversification.

The Indo-US joint venture

The Indo-US joint venture was an attempt by both partners to improve their own standing in global markets. For the Indian partner, the joint venture was part of the diversification program, while for the US partner, establishing a presence in India would provide a cheaper manufacturing source. Although the US firm usually established greenfield ventures as wholly-owned subsidiaries, an exception was made in the case of India given the limited knowledge of the local environment. The joint venture (Bearings, Inc.) was thus established to manufacture roller bearings for truck and tractor wheels, transmissions and gear boxes, diesel engines, and other general engineering applications. In addition to standard bearings, a special cartridge bearing was also to be manufactured for use in railway and industrial applications.

Bearings Inc., had an annual turnover of INR790 million (US$23 million),[8] a paid-up capital of INR424 million (US$12.5 million), and 450 employees. The US partner supplied the plant and technology, consisting of imported reconditioned equipment restored to have a normal life equivalent to new machines. The reconditioned equipment was also retrofitted with state-of-the-art electronic controls, in-process gauging and other proprietary apparatus. Bearings, Inc. had the right to use the US parent's technology, including know-how and other proprietary information relating to design and product specifications and methods of manufacture for a period of six years. According to the terms of the agreement, the Indian company could market these products in Sri Lanka, Bangladesh, Nepal, and Bhutan.

The technical, marketing, and administrative leadership of the new venture was provided by managers from the United States and Europe while the personnel and finance functions were handled by managers from the steel or bearings subsidiary of the Indian partner. All other full-time employees were either on deputation from the Indian parent or direct hires for the venture. In addition, the US parent also provided technical expertise and training through personnel who were sent for short assignments from overseas locations. Implementation of initial phases occurred in stages during 1989–95; however, external factors such as recessionary conditions in the automobile industry, lack of orders from the Indian railways, increased import costs (a consequence of the devaluation of the Indian rupee) contributed to project delays, and costs escalated dramatically. This prompted efforts to seek a reappraisal of the project cost by the Industrial Development Bank of India (in financial year 1994–95) to finance the project and increase its licensed capacity.

Bearings Inc. was largely managed by the US parent through two groups, one based in the United States and the other in its European offices in the United Kingdom and France. The European group took the lead in developing the project plans for the Indian plant. The Indian plant was to be designed as the most sophisticated facility technologically from both the equipment and work-flow standpoint. With the assistance of a management consultant, the European team conceived a plan to maximize flexibility in manufacturing operations. Short-cycle manufacturing concepts were incorporated to minimize work-in-process, safety stocks, and cycle times in the Indian plant.

Although the US and European teams recognized that the work environment in India was very different, their understanding was limited and not based on experience. Hence, while the decision to establish the venture was taken in 1987, by the time plant construction began, it was January 1990. Machinery installation was also delayed and began in 1991. By late 1994, production was still at only 75 percent to 80 percent of capacity. Changes in management also affected the continuity and cohesion of the project. The core group first assigned to the Indian venture in September 1989 was replaced in 1992–93, resulting in a change in management style. Financial problems also complicated the situation.

Despite these difficulties, a quantum increase in production was achieved in the last quarter of 1994 (monthly production increased from 70 000–80 000 bearings to 120 000 bearings). This increase in productivity was achieved after top management initiated a dialogue with employees to apprise them of the firm's financial position. This information prompted organizational members to take personal responsibility and work towards improving the situation. As of the end of fiscal year 1994–95, the joint venture had reduced its losses and profits were projected for the following financial year. The break-even point was attained in 1995–96 but at double the capacity originally projected.

The technology transfer experience in Bearings differed considerably from that of Earthmovers. Although external factors such as devaluation, relationships with suppliers, and a recessionary market affected project deadlines and cost, other organizational factors associated with project management contributed to slow knowledge assimilation and absorption. The pace of development of technological capabilities was much slower than expected by the partners. Delays also increased project costs, impaired the firm's financial position, and reduced credibility with suppliers. The dynamics of the transfer process are examined below.

Fostering relationships The supplier–recipient relationship in the Bearings joint venture inhibited the creation of a community of practice. At the outset, the US partner, as expert, controlled the planning process by forming teams in Europe and the United States. Since the Indian partner's involvement was not sought, US directors and experts were unable to incorporate critical local information vital to manufacturing and marketing in India. Local non-involvement also contributed to expatriate managers' perception that expertise was located overseas rather than in India. The manufacturing director noted: 'The production engineers, the supervisors, the operators, they just wait for the foreigners to come and give them the update on what they should do'.

Communication difficulties arising from reliance on foreign experts (32 percent compared with 14 percent in Earthmovers and 15.25 percent in Steelworks)[9] and imported equipment resulted in longer lead times for new product introductions, delays in solving minor problems, reduced productivity, and extremely high communication costs.

Second, learners were denied access to full information on the grounds that technology supplied was proprietary. Hence, on-site problem-solving was difficult because relevant manuals were not available to engineers. A maintenance engineer commented:

> The company [foreign partner] thinks that there are a lot of things that are proprietary. [Such knowledge] may be proprietary, but it may be an essential thing to run the plant . . . because unless you know the way they arrived at a process or a

> decision, we cannot go back and question and go on a different path to maybe arrive at a better decision. . . . there have to be things that are proprietary – anything and everything cannot be handed down – but where knowledge is the main thing, for example, in maintenance, that's where a person should know how to do things. In such areas which involve knowledge, transfer of knowledge must be complete.

Expatriate managers noted that inter-departmental relations needed improvement, and commented on the prevalence of hierarchical attitudes and attachment to status and titles. The sense of being part of a young, elite group generated competition and information hoarding amongst peers:

> We've hired people who are very literate, young people full of energy and sometimes that doesn't go together. These guys are very arrogant and the management style is very directive. There is a lot of competition between them because they're young and very ambitious . . . and they fight each other. There's very little teamwork in our group.

Management–worker relations were egalitarian as exemplified by the accessibility of top management. However, unlike the other two firms, Bearings did not initially recognize the need to establish relations with local suppliers, since all plant equipment was imported from the United States. One of the directors acknowledged that:

> It's [the plant] been designed by European engineers for Indian conditions, so it doesn't work . . . there are some things such as plant engineering, that when I look at what our Indian partner does, they do it much more cleverly than we do . . . they use local suppliers, they've got good knowledge of local conditions; maybe the things are not of the same standard and quality, like the air-conditioning unit. They've got local service and we haven't, because we have chosen American companies. We have suppliers, the worst case is that the suppliers have an agency in India; so we buy from the Indian agent, but he's importing it from the US or from Europe . . . Most of our suppliers are on the West coast. That's one of the problems. There are local suppliers, our partner has used them, another Indo-American joint venture has used them; you don't have to go to the West or the South for some of the things.

Organizational design for knowledge transfer Lacking local contextual knowledge, the Bearings planning team in Europe and the United States was unable to foresee the heightened need for coordination with local suppliers, coordination between project groups located in Europe, the United States, and India, and coordination between different functional departments. Lack of coordination resulted in delays from unanticipated differences in work methods. A technology director contrasted local cross-functional coordination unfavorably with that in the French and British plants of the parent company:

> It's very functional in India. Everybody's got his own empire . . . You expect for a plant of this size (about 400 people) to have much more integration. I won't compare with the British (1800 people) and French (800 people) plants where really you have functional heads with a lot of people; but it works sometimes better than over here.

He also noted that while the French plant had one supervisor for every 22 people, the Indian plant had one for every seven.

Overall, the level of coordination was not very high. Comparison of the integration index in Bearings (derived from responses to 20 questions on project organization)[10] with that of Earthmovers suggests that the project was not as highly integrated as the latter. In Bearings, the index is 0.52 or 20 percent lower than that observed at Earthmovers (0.72).

Communication difficulties stemming from differences in the work environment and infrastructure also hampered coordination. The local work environment created difficulties for a management cadre that took state-of-the-art communication for granted:

> Between the UK and the US or Europe, you sit at your screen and type out your message yourself and next morning it's replied to – it all comes over the network. You can't do that here; working with courier services, the delay, the telephones . . . If we had the same sort of communication that exists between the US and Europe, that would be a tremendous help here (e-mail etc).

Dependence on assistance from overseas and the involvement of groups located in Europe, the United States and India with different priorities and concerns exacerbated the situation and made communication a critical issue. According to a senior manager:

> It was more of a mindset than a magnitude of communication problem. The reason why it became such a big problem was due to the fact that local management was totally dependent on the US or UK for every decision. If most of the decision-making had been in India, then it would not have been a problem.

Opportunities to practice In Bearings, interactions with experts were limited, owing to geographical distance and the inadequacy of local expertise. Fifty Indian managers were trained in the United States but were unable to train operators on their return because they received training in rebuild and manufacturing rather than in production. Thus, training was 'not very effective' according to one of the senior executives:

> You really don't learn much as an outsider in machine rebuild. Where you do learn, is out on the shop floor, on the production floor . . . because on the production floor, you experience all of the factors that affect production, including

> if the machine is down and requires machine repair and the maintenance people come in; you can see the dynamics of it; these people were not exposed to that.

Training was not synchronized with the arrival of equipment from the United States. The differing priorities of managers in Europe and the United States also contributed to delays in the dissemination of drawings, specifications, and new product introductions, affecting the speed of transfer. Eventually, solutions included accessing technical assistance via weekly teleconferences and bringing in experts from the United States.

Mastery: evidence of knowledge transfer Learning in Bearings was initially inhibited. Competence development was limited to areas where local employees were able to initiate problem-solving and were able to engage in dialogue with experts. In the heat treatment area, for example, performance improvements were observed because batch integral quenching furnaces were used for the first time in the Indian plant instead of the pit furnaces prevalent in overseas plants. Hence, local managers had greater autonomy in problem-solving and evolved solutions through trial and error. By early 1994, rejection rates in the heat treatment area were reduced to 4 percent from about 40–50 percent in June 1993.

Despite initial difficulties reflected in poor financial performance, management's attempts to create a community of practice initiated learning in 1994. Plant output improved by 50 percent in the six months following the period March 1994 to September 1994. By 1995–96, Bearings also made headway in reducing cycle time, work-in-process, and inventories. While indigenization reduced the cost of new equipment and spare parts by 40–50 percent, dependence on the foreign partner continued because of low volumes. Quality certification for international markets was obtained in early 1995 and by 1995–96 the quality of bearings produced in India was equivalent to that of other overseas plants (as established through periodic testing overseas). By 1995–96, Bearings was beginning to enter global markets through exports. However, the plant did not yet have the production, maintenance, and design capability required for autonomous functioning.

In summary, although the intention was to encourage the development of both objective and subjective knowledge, the initial inability to create a community of practice resulted in inhibited learning and slower transfer.

Technology Transfer in a Steel Company: Apprenticeship for Strategic Renewal

Steelwork, established in 1807, has an integrated steel plant and raw materials are obtained from its own mines and collieries. It is the Indian partner described earlier in the Indo-US joint venture.

Modernization was critical for enhancing the company's competitiveness and productivity to enable it to withstand competition from substitute materials and new technologically superior plants. A modernization program was initiated in the 1980s, primarily to replace antiquated equipment.[11] An integral part of this plan was the installation of a state-of-the-art blast furnace to increase capacity and productivity of hot metal production (the new furnace had a capacity of approximately 1 million tonnes per annum, and could produce about one-third of the total volume of hot metal currently produced by the steel plant). Steelworks' senior management took the decision to purchase a blast furnace manufactured in Italy for use in Portugal in 1981, but never installed in Europe because of overcapacity in the European steel industry. As a result Steelworks was able to acquire it at a total cost of DM200 million (the cost of the furnace was about one-fourth the cost of a new one; for example, a new one with a capacity of 10 000 tonnes per day was erected in Germany at a cost of DM800 million). The US$100 million project was expected to increase crude steel production capacity to 3mtpa[12] and raise saleable steel capacity from 2.1 mtpa to 2.7 mtpa. In comparison with the most recently commissioned furnaces (1958) the new furnace was highly automated and had other advanced features.

The detailed engineering, project management, and supervision were assigned to a special group responsible for various modernization projects; the group was free to consult experts when required. Consultants specializing in the steel industry were used (including a leading Indian consulting firm dedicated to the steel industry, the Italian manufacturer of the blast furnace and a leading German steel manufacturer). Technical discussions, interactions, and visits to operating plants were part of this program.

Site construction began in 1989–90; in late 1991, a divisional manager was selected to lead the project and other team members were recruited from various areas to begin training and commissioning the furnace. The furnace was stabilized rapidly despite unexpected problems encountered in commissioning and other delays. Although the furnace was finally 'blown in' in November 1992, eight months behind schedule, it reached its rated production of 2700 tonnes on the fifteenth day and achieved a coke rate of 600 kg/tHM[13] by the twentieth day, achieving both targets much faster than expected. Members of the blast furnace team spoke of the project with a sense of pride and achievement and confidence that the technological sophistication of 'their' blast furnace compared favorably with the best furnaces worldwide.

Fostering relationships Rather than entering a technical collaboration or forming a joint venture, managers in Steelworks attempted to fashion a community of practice independently, using technical consultants and a variety of local suppliers instead of relying on a single expert supplier. Direct communication between internal departments and technical consultants was achieved through a special task force set up to speed project implementation. Despite strong links with suppliers and consultants, project members recognized that they were ultimately responsible for results. The head of the modernization group commented: 'The consultant is only a consultant and while he is definitely responsible for the project completion, we cannot hold him accountable for any delays'. He also acknowledged the assistance of the foreign supplier: 'We really got a lot of help from the German firm. A lot of systems were new to us . . . the people in the German firm had built up this thing over a period of 18 years. We tried to pick up everything'.

Inter-departmental relations in Steelworks were cohesive owing to members' awareness of how critical the project was for the organization: 'Every month, there was so much of interest loss, so much of depreciation, so much of lack of production, so much loss of face in the market . . . the firm was going through a tough time'.

A paging system facilitated instant communication when problems were discovered. Finally, the need for self-reliance in the face of stringent targets compelled project members to improvise solutions and led to the formation of a 'self-organizing team' in the sense of Imai et al. (1985, pp. 337–75). Management–labor relations also focused on completing the task on time by supporting workers. A manager commented on the high levels of motivation in the team: 'We have had quite a few occasions when people felt unfit, they were hospitalized and they wanted to come back from the hospital as quickly as possible to join here . . . That is the kind of team spirit we had'. However, local suppliers were characterized as unable to adhere to delivery schedules, making project management difficult.

Organizational design for knowledge transfer As in Earthmovers, a fairly high degree of concurrence and organizational coordination was also achieved in Steelworks. Parallel engineering was included as part of the project design since all stages of commissioning had to be completed before blowing in the furnace. Coordination was also achieved through scheduled meetings to discuss current problems, structured training sessions, and by rotating members through different functional areas. A project member observed that he had learned much about project management, including:

> For example, how to think proactively about what you will need next. Let's say, for example, many times in this project, we reach a certain stage and then we find that we don't have the equipment needed, and then we go for designing that equipment. So if you are able to foresee all those stages, and the equipment needed for each stage, it helps. It's not done serially, most of the jobs are done in parallel.

Opportunities to practice Over 50 managers and operators were provided training in India (core group members were sent overseas) before the furnace was commissioned Project members acquired theoretical knowledge by studying relevant literature and practical knowledge by developing scale models at the R&D center and improvising experiments during the early stages of commissioning the furnace. For example, by tripping the hydraulic system, engineers determined how long each part would take to revive. Involvement in practice was reinforced by the decision to learn every phase of project implementation. A divisional manager commented: 'We installed a lot of instrumentation; during the installation our philosophy was that if we do it ourselves, there will be nothing like it because then you have had the first-hand experience of doing the things, and so, maintenance would be very easy'.

Mastery: evidence of knowledge transfer Mastery in Steelworks is evident in the ability of project members to solve problems innovatively during critical stages of technology implementation. Examples include handling difficult welding problems on furnace structurals and heating the blast furnace stove without reliance on external technical experts. Stabilization of the project took only 15 days. Moreover, team members reported that they did not 'just copy' what they had seen but had built in additional flexibility into a system with a lot of limitations by modifying specific areas. Thus, performance-wise, the availability of the furnace at 95 percent was comparable with the best in the world.

As in Earthmovers, data-gathering and a problem-solving orientation were regarded as critical. The effort to raise quality throughout the organization dictated training through seminars and workshops led by experts (e.g., Dr. Taguchi, a quality expert). Core members of the team who moved out of the project actively disseminated knowledge and acted as a catalyst for change by replicating communities of practice in other areas of the firm. (For a summary of findings from the three focal firms for Stage I [and II], see Table 4.1.)

Other Indian Firms

Evidence from seven other firms supports the conclusion that the transfer process must use appropriate organizational systems to ensure knowledge assimilation, dissemination, and the creation of firm capabilities. Capability creation is an inevitable consequence of exposure to new technology if knowledge transfer occurs. However, the amount and extent of learning is constrained by the image organizational members have of future possibilities and the time and attention devoted to preparing for the detailed work associated with introducing new technology. Firms are more similar than different, for several reasons. First, members of all firms interviewed viewed themselves as part of an elite group, the leading firms in the industry with links to the top global multinationals. Access to ideas and models of organization prevalent in these firms enabled them to adopt industry best practices. Second, economic liberalization and enhanced participation by foreign firms increased competition for domestic firms. Indian firms entered a transition phase, moving from operating in a milieu of scarcity in which self-sufficiency was important (for example, in the pre-liberalization era, import substitution was mandatory to maintain a stance of independence), to a situation where it was necessary to scan the world for the best and cheapest inputs to compete successfully in global markets. Although leading Indian firms had begun to foster learning (for example, by working with local suppliers and other relevant members of the community of practice) attaining mastery would take time. While technology absorption can, theoretically, be rapidly achieved, it is constrained by the amount and quality of managerial resources and how effectively these are deployed.

A summary of key findings from seven other organizations based on the analytical framework applied to the three focal firms is presented below. Interviewees were managers and executives of firms in the engineering sector headquartered in three different geographical locations in India – Delhi in the North, Mumbai and the Western industrial belt, and Bangalore in the South. The sample included firms from the agricultural equipment sector, automotive components such as pistons and rings, power equipment, motorcycles, utility vehicles, tractors, automobiles, and construction equipment. In four of these companies, the transfer mode was technical collaboration (products included engines, automotive components, tractors, and hydraulic excavators); the other three firms transferred technology via joint ventures (industry sectors included were power equipment, motorcycles, and automobiles).

Fostering relationships Managers interviewed confirmed that the relationship with the technology supplier was vital to successful technology transfer. Most reported good relations with the supplier; however, managers in two firms indicated that their relationship with the technology supplier had virtually ceased, partly because these were old technology collaborations that were being re-evaluated. In one case the French partner had supplied the Indian firm with technology for utility vehicle engines and was now considering an alliance with another Indian firm for manufacturing automobiles. In the second case, the Indian firm was considering alternative options such as in-house development, technical collaboration or joint venture for upgrading technology to match domestic rivals' alliances with the leading Japanese firms.

During the technology transfer process, interviewees in most firms recalled that relations with suppliers were excellent. Little day-to-day involvement was required in technical collaborations and recipient firms were usually specific about what they required from suppliers. For example, the technical director of a leading pistons and rings manufacturer noted that:

> We often invite our foreign collaborators to visit us and to see what the facilities are here and to discuss together and identify what are the gaps . . . It sometimes means buying new machinery from them or a new type of technology to manufacture. . . . So, after deciding what the gaps are, we [determine] how these should be filled with new equipment or manufacturing process . . . and then we see how to transfer that here.

Such interactions also helped the technology supplier to understand more clearly what would be suitable for the Indian environment and to assess the recipient's current level of technical capability. The director noted that on account of the rapport established, the Japanese collaborator had stopped a production line for ten minutes to explain how to dismantle and assemble for a line change while they were visiting the plant in Japan and also allowed them to take photographs. Similarly, the head of the agricultural division of a conglomerate manufacturing farm equipment, motorcycles, and construction equipment also echoed the importance of rapport between the supplier and the recipient: 'It is the degree of the relationship that determines the extent to which you are open and what advantage you take'. Moreover, he recognized that relationship development was idiosyncratic and depended on the personalities of top managers:

> It is the view of top management which is very important. . . . We have always insisted on having one or two on our team from our collaborators. Thus, they become a very important link for us . . . I can think of a person, a guy everybody knows . . . whether you have a product problem or a marketing problem or any type of problem, he can ring up the concerned guy because he knows how an organization works. . . . Communication is tremendously facilitated by virtue of the presence of foreign technicians, but the extent to which it is successful depends upon the individual or individuals. There have been cases where people

> were frankly not interested . . . and nothing happens. You had a different guy who was enthusiastic and knew everybody in the organization and the two organizations became much closer.

Relationship building also helped to improve receptivity to new ideas. However, sometimes collaborators did not share know-how as readily in areas such as maintenance, according to engineers at a pistons and rings manufacturing company with a Japanese collaboration for valve technology. Lack of knowledge-sharing in this instance was attributed to a different philosophy emphasizing replacement rather than maintenance. Alternatively, concern about knowledge appropriation by the Indian partner may have been another reason for lack of knowledge-sharing. An interviewee noted: 'So far they have not allowed any of our maintenance engineers to be trained in their maintenance shop. . . . Maybe it is sacred . . . if we know [how to do] maintenance, then we can do it ourselves'. Consequently, while relations were generally open, commercial decisions on the part of both the supplier and the recipient influenced certain interactions.

Relationships with domestic suppliers were also in a state of transition. In the decades before liberalization when local competition was minimal, and suppliers could sell all they produced, there was little customer orientation and attention to quality. The concept of educating suppliers was adopted by many of the companies examined in this study. In the majority of the firms studied, steps were being taken to develop the suppliers' knowledge base by providing them with technical information and process know-how. In addition, many firms had also formed supplier development departments. They also emphasized the importance of suppliers in providing the infrastructure for effective absorption and assimilation of technology. For example, managers in a utility vehicle manufacturer that had a technical collaboration for engine manufacture with a French automotive company paid special attention to educating suppliers. In addition to the training provided by the French partner, four people were deputed from France to conduct a detailed inspection of vendors, which included seeing how the component was machined and tested. Other interviewees mentioned learning from the supplier relationships of a leading Indo-Japanese automotive manufacturer. Thus, the supplier selection process and ongoing interactions served to transmit know-how to and involve a larger group of firms than the initial technology recipient, suggesting that the rate of development of members of the wider community is likely to have an important effect on the rate of technology absorption of the importing firm.

Finally, intra-firm relationships were largely characterized as good. In most firms open communication across departments was encouraged via suggestion schemes. Sports and family entertainment activities helped to cement social relations. In the Indo-Korean automobile joint venture, the layout of administrative offices was designed to promote increased communication between different functional areas. While engineers/managers occupied a central open office and senior managers were allotted separate cabins, these rooms were separated from the central office by glass panels to emphasize transparency. However, while relations between management and workers were cooperative[14] and efforts were made to build trust, inter-departmental relations needed improvement. Although most firms were aware of the need to promote inter-departmental communication, in some cases, transparency had not been achieved. For example, in the Indo-French construction equipment collaboration, a manager commented that the R&D department had kept itself aloof from manufacturing and considered itself the 'brain of the company', was not willing to become part of problem-solving, and had adopted a 'me versus you' attitude, which was contrary to efforts to encourage collaboration. Another manager in the same organization noted that while 80 percent of the people were accessible, some individuals had feudal attitudes and would not take readily to open communication and questioning assumptions.

Organizational design for knowledge transfer In all firms, the model of concurrent organization was fully accepted by managers and new projects were designed accordingly. Moreover, projects begun before liberalization in the mid-1980s were restructured to facilitate concurrent processing of tasks. In the Indo-French collaboration for engines, the R&D head reflected that if he had to do things over:

> As an R&D man, I would have got R&D involved in it from the beginning so that know-how and know-why absorption would have been better . . . Today in the liberalized atmosphere, some of the decisions would be different. At that time we had to achieve what we set out to do within a lot of constraints.

Nevertheless, two plants had been reorganized from 1994–95 (reorganization at one of the plants had resulted from managers' effort to examine all processes while operating the plant during a labor strike). In the construction equipment firm a manager noted that coordination was difficult to achieve because: 'In Indian organizations in many cases we are not able to move forward because we are confronted by egos, level consciousness, and by wrong symbols of power and fame'. However, increased competition and a falling market share had triggered a process of self-diagnosis and evaluation of their current position, resulting in an effort to achieve greater integration across functional lines.

Information technology was also viewed as an important way to establish new linkages across departments and accelerate information transmission. While some organizations were investigating network options, the main mode of communication – in addition to face-to-face interactions – was via telephone, fax machines, and the use of courier services in place of the postal system. In the two Indo-Japanese and the Indo-Korean joint ventures, coordination was emphasized from the start. In all three cases, there was a high degree of involvement from the supplier's side. In the Indo-Korean automotive joint venture, the entire manufacturing organization had been replicated in India. However, despite recognition of the need for coordination, formal hierarchy was not abandoned. Coordination was largely achieved through the formation of problem-solving groups, cross-functional teams, and meetings. Also, junior plant engineers had access to top levels of management for problems that could not be resolved at lower levels.

Opportunities to practice The commitment to learning and attaining both 'know-how' and 'know-why' were demonstrated in all firms examined. The need for inculcating new skills was well recognized and most senior managers were convinced that employees at all levels needed exposure to new techniques and training in new skills. At the operator level, training was particularly important if operators were expected to work with more advanced technology and be responsible for delivering high-quality products. Top management was aware that direct experience at the shop floor level was important. An executive in the pistons and rings firm observed that: 'Unless you expose a reasonable number of people to the practices being followed elsewhere they will not follow . . . when it comes to basic machine operation, shop floor people must interact with the corresponding shop floor people. That is the only way; no other way will work'.

However, there were variations in (1) the amount and kind of training imparted, (2) initiatives taken by the recipient firm in the attempt to make the technology operational in the new environment, and (3) the cooperation and support provided by the technology supplier. The extent of exposure varied: in the Indo-Korean joint venture 100 percent of the employees were sent for training to Korea, whereas in the Indo-French technical collaboration for engines where a small group of managers visited the French collaborator's facilities for four–six weeks with follow-up visits as the need arose. In the Indo-French collaboration, overseas experts were also sent for two or three months to provide technical assistance in India to help establish an aluminum foundry. Thus, the idea that employees should be empowered to widen the knowledge base had gained widespread acceptance.

In addition, workers were often provided training in other skills besides technical skills, including training in communication and collaboration. Lack of codification rendered on-site training essential. The technology director at the pistons and rings firm noted that: 'Theoretically speaking, yes there are documents, there are managers to explain why it should or should not happen, but . . . you will find that it will take a longer time, there will be many pitfalls, many things you will not be able to figure out'.

However, despite collaborative efforts, most plant engineers noted that transfer was not systematic and that successful production was largely a result of their own efforts. Moreover, they exhibited confidence in their own manufacturing capabilities, a reaction reminiscent of the findings of Westphal et al. (1985) in their study of technology transfer into Korean firms. In some cases, firms had absorbed sufficient knowledge about the change process to diagnose what was relevant for their situation. Overall, training was considered important and members were provided opportunities for learning.

Mastery: evidence of knowledge transfer Indications of progression towards mastery and objective knowledge transfer are evident in improvements in design capability and performance. In most firms, interviewees displayed a sense of pride and ownership of the technology they had acquired and, in some instances, even claimed that they had improved on what they had learned. Interviews revealed that in six of the seven firms, design capabilities were limited. While most firms were proficient in absorbing information about specifications and procedures provided by the supplier and using this knowledge for production activities, the majority indicated their inability to initiate technological change. Among reasons cited for lack of innovation in the design area were the lack of infrastructure for R&D in comparison with firms in industrialized countries (the construction equipment manufacturer noted that in comparison with a US competitor's R&D spending of 3.5 percent of a turnover close to US$1 billion (i.e., US$35 million), its own spending amounted to a miniscule 1 percent of INR3 crores (less than US$1 million) on a turnover of INR300 crores (less than US$100 million).

Similarly, an R&D manager in the engine manufacturing firm who compared the strength of its R&D department with that of a Korean manufacturer of engines indicated that the former had only five engineers in its R&D departments whereas the latter had 500 engineers its engine group. Also, designs were generally taken as given since they represented tried and tested technology. Innovation in the design area was dictated by local demand conditions (the need to produce more basic, multifunctional machines since the market for specific-purpose machines was limited) in the pre-liberalization era. Examples of innovations included modifications to prevent the engine from overheating because of higher ambient temperatures or the modification of a tractor design to make it suitable for wet paddy cultivation. However, to improve the 'ergonomics of the tractor' to meet international standards of operator comfort they would have to source technology from overseas. Reliance on foreign expertise seemed inevitable even when indigenously designed equipment was available because of the need to meet global competition. Also, most R&D departments were in a state of transition, in the process of either being established or upgraded and managers acknowledged that they had learned a lot from suppliers:

> In the French designs, most of the information is fantastic. They give all the specifications and parameters on drawings; they are well defined. Now in the French way of working, they go one step ahead. This drawing is again analyzed by the quality control department . . . [and quality standards are incorporated into the design].

Although global competition, lack of local knowledge and the poor image of Indian products were cited as reasons for reliance on foreign expertise, a manager suggested an alternative perspective. Although Indian engineers were very knowledgeable and highly competent, lack of capability to implement technological change stemmed, in his view, from lack of confidence. He added that: 'We don't understand design in its overall aspect. This is a problem in India because of the emphasis on import substitution. Senior management has probably not looked at technology as a whole'. Also, R&D departments were more focused on testing rather than on designing components and products.

At a subjective level, knowledge transfer resulted in the development of an emphasis on quality, productivity, and customer service in most firms. The engine manufacturing firm achieved a productivity improvement of 250 percent after a period of stagnation and labor unrest in mid-1994. The generation of large amounts of data as a consequence of increasing automation both on the shop floor and the use of information technology for data collection from the field dictated using more rigorous problem-solving methods. Engineers at the pistons and rings manufacturing firm expressed a need to learn more about methods for data analysis such as statistical process control. Many other firms were already using such methods, indicating a trend towards greater reliance on objective data for decision-making. At the construction equipment plant a comprehensive four-tiered quality system was implemented, including documentation in a quality-system manual, evolving procedures, developing instructions and maintaining records. For example, in the case of a weld failure reported by a customer, the problem was first defined, and data collected and analyzed using quality improvement tools such as histograms, Pareto diagrams, and cause and effect diagrams. Corrective action involved either manufacturing changes, design changes, or changes in raw materials. Measures were also identified to evaluate the effectiveness of the corrective action taken. Thus, elements of new social and administrative systems were being incorporated along with new technical systems.[15]

Evidence suggests that Indian firms were beginning to embark on a process that would lead them to becoming global participants by increasing their awareness of worldwide markets and the links between domestic and foreign markets. However, these firms were as yet peripheral participants in the global economy.

STAGE II: INSTITUTIONALIZING LEARNING

A second round of 12 interviews in March–April 1999 suggests that all three firms adopted a more proactive stance towards learning by initiating changes in identity to match prototypical global manufacturing firms.

Earthmovers

On 1 April 1999, Earthmovers became an autonomous organization engaged in establishing a new state-of-the-art mini-excavator plant closer to customers in South India. It also entered a joint venture with the Japanese supplier taking a 20 percent stake in the new company and pursued other foreign collaborations to meet customer demand for product variety. Design adaptation continued with upgrades of the earlier line of excavators to meet customer demand for new applications and ease of operation. CAD (computer-aided design) systems were used to enhance productivity in design and to access to design specifications online from technology suppliers. However, research engineers reported that pressure to generate new products and to reduce product development cycles resulted in increased dependence on foreign R&D while slow infrastructure development continued to constrain growth.

Bearings

On 1 April 1999, the joint venture was dissolved and Bearings became a wholly-owned subsidiary of its US parent. Subsequently, top management focused on developing an autonomous operation and pursued productivity and quality improvements via QS 9000 certification. Fifty percent of the product was exported to the United States and Europe in 1997–98, and the Indian subsidiary also provided assistance in establishing a new plant in China. Although Bearings began to develop strong relationships with local suppliers, new product introduction, still controlled from the United States, remained problematic. Finally, although R&D capability in the manufacturing subsidiary remained adaptive, the US parent established one of its five global research centers in Bangalore, India, to access local capabilities.

Steelworks

Assailed by over-capacity in the industry and competition, Steelworks embarked on a major new project (commissioned in April 2000) to produce value-added cold rolled steel for the automobile and white goods industry. Prior experience with installing the new blast furnace enabled managers from Steelworks to direct the project. Thus, project leaders scanned world markets for specialized knowledge for producing cold rolled steel and the project management and technical expertise of leading Japanese and Korean steel manufacturers. Project organization was integrated and used a concurrent processing pattern with intensive communication and coordination. The speed and cost of all activities were benchmarked against the world's best standards. The project was completed ahead of schedule and surpassed the world record of 28 months by over one-and-a-half months. New human resource practices such as attitudinal (psychometric) testing for employee selection, and group incentives were introduced and the organization restructured, reducing the number of hierarchical levels to foster the diffusion of knowledge.

CONCLUSION

The findings support an evolutionary framework of situated learning and suggest that knowledge transfer was facilitated through the formation of cross-border communities of practice that enabled both the growth of objective knowledge of manufacturing and product innovation, and of subjective knowledge via an identity shift towards membership in an international community. However, learning to become global is a non-linear process fraught with difficulties. As the Bearings case shows, knowledge transfer was initially impaired, because of the local partner's lack of legitimacy, poor access, and inadequate participation, which hindered the formation of a community of practice. In contrast, strong communities were established in both Earthmovers and Steelworks, fostering access to know-how and promoting practice and autonomous experimentation in these firms.

Table 4.1 Learning via technology transfer – Stages I and II (three focal firms)

Framework Category Community Creation	Earthmovers	Bearings	Steelworks
Stage I Fostering relationships:			
a. Supplier–recipient relations	a. Supplier–recipient relations were collaborative, characterized by information-sharing and mentoring	a. Supplier–recipient relations were initially characterized by lack of trust, little information-sharing and mentoring and low involvement of local partner in initial planning and technology implementation	a. Supplier–recipient relations that were formed with global industry experts and consultants were collaborative with the recipient initiating the interactions
b. Inter-departmental relations	b. Inter-departmental communication was strong with emphasis on improving working relations between departments via inter-departmental dialogue	b. Inter-departmental communication was high but the level of expertise available was low because of inexperienced hires	b. Inter-departmental information sharing & communication & joint problem solving were emphasized
c. Management–worker relations	c. Management worker relations focused on human development, providing meaning in work, & on workers' personal needs to nurture commitment	c. Management worker relations were initially biased; greater trust in relations after 1994	c. Management worker relations promoted development through experimentation
d. Recipient–local suppliers relations	d. Recognition of need to develop local suppliers to increase volume without sacrificing quality	d. Local supplier development began by 1996	d. Local suppliers regarded by planning as not within control, especially where meeting schedules was concerned

Organizational design for knowledge transfer:			
a. Coordination/ integration	a. Coordination/communication viewed as critical for modular design of hydraulic machines	a. Difficulties in communicating & coordinating with groups in India, United States and Europe	a. Coordination achieved through adoption of a team structure and through rotation of members through different functional areas
b. Concurrent processing	b. Concurrent task processing achieved via cross functional groups, small group activity, E-mail and other computerized systems	b. Cross-functional coordination limited and little teamwork. Low level of integration; more hierarchical organization than in similar larger plant in France	b. Project designed for concurrent processing but difficult to implement because of reliance on local suppliers
Opportunities to practice:			
a. Training and practice	a. Eighteen welders sent for training to supplier's plant for six months in Japan; Japanese experts also provided training in India on welding practices, jigs & fixtures, plant layout, production, etc. Indian engineers also obtained information through industry fairs and conferences	a. Training not well coordinated, indirectly useful, & of insufficient duration; training initially provided to managers rather than production workers	a. Over 50 managers and operators were provided training. Theoretical knowledge developed through literature and practical understanding through scale models
b. Associated knowledge	b. Moderate knowledge of associated functional areas	b. Knowledge of associated functions lower than average	b. High level of associated knowledge. Learning emphasized throughout implementation. Reliance on technical experts was minimal

Table 4.1 (continued)

Framework Category Community Creation	Earthmovers	Bearings	Steelworks
Mastery: evidence of knowledge transfer	*Objective knowledge:* Design capabilities upgraded; reduction in cycle time in heat treatment operations by 50%; little reliance on foreign technical experts; use of new problem solving techniques and small group problem-solving; continuous improvement measures implemented via feedback seeking, diagnosis, testing and evaluation for every activity; ISO quality certification achieved	*Objective knowledge:* Competence development was limited to areas such as heat treatment where the foreign partner's knowledge was limited; rejection rates in heat treatment dropped to 4% from about 40–50%; by 1995–96, cycle time, work in process and inventories were reduced; some indigenization took place but dependence on the foreign partner continued. ISO quality certification obtained in 1995.	*Objective knowledge:* Improved technical problem-solving capability; handled difficult problems during implementation without reliance on foreign technical experts, e.g. welding and heating; blast furnace performance and 95% availability was comparable to that of the best blast furnaces worldwide.
	Subjective knowledge: Increased reliance on quantitative data; customer and market orientation; new identity as a top quality, world class manufacturer of earth moving equipment; language of analytical problem	*Subjective knowledge:* Perceived as a manufacturer of high quality bearings; by 1995–96, the quality of bearings produced inIndia was equivalent to that of other overseas plants; exports to global markets began.	*Subjective knowledge:* Data gathering and problem solving orientation emerged as a result of the transfer; quality orientation was also enhanced; efforts were also being made to inculcate a market orientation; new identity of world class steel

	solving and continuous improvement permeated the organization		maker emerged; new language used to describe organization, i.e. 'a learning organization'
Stage II Institutionalizing learning	Earthmovers was spun out as an independent unit in 1999 and the Japanese collaborator took a 20% stake in the new venture; adaptation continued in design and manufacturing and new computerized systems were implemented	In 1999 the joint venture was dissolved & Bearings became a subsidiary of the US parent; QS 9000 certification was obtained and 50% of the product was exported to United States and Europe; the Indian subsidiary assisted in establishing a new plant in China; development of local suppliers but local R&D capability remained low	Learning was diffused to older blast furnaces; a new project to manufacture cold rolled steel in 2000 leveraged knowledge from the blast furnace project; project organization was integrated and used a concurrent processing design; the project was completed in 28 months surpassing the world record; new HR practices introduced & language disseminated

*Table 4.2 Learning via technology transfer – Stages I and II (seven other manufacturing firms)**

	Firm 1	Firm 2	Firm 3	Firm 4	Firm 5	Firm 6	Firm 7
Industry	Auto components	Power equipment	Agricultural equipment	Transportation – motorcycles	Automotives	Utility vehicles	Construction equipment
Type of collaboration; (year established)	Technical collaboration; (1975; 1985; 1991 onwards); licensee (1985)	Joint venture (1985 onwards)	Technical (1969); consultancy (1980s), joint venture (1990s)	Joint venture (1984 onwards)	Subsidiary of a foreign MNC (1992)	Technical collaboration (1984)	Technical collaboration (1975)
Stage I Fostering relationships:							
a. Supplier-recipient relations	a. Good but insufficient knowledge sharing	a. Good	a. Frought with control issues	a. Good	a. Good	a. Excellent	a. Good but insufficient knowledge sharing
b. Inter-departmental relations;	b. Increasing Collaboration	b. Good	b. Hierarchical	b. Good	b. Good	b. Improving	b. Not good; improving
c. Management–worker relations	c. Good	c. Good	c. Hierarchical	c. Good	c. Good	c. Improving	c. Good
d. Recipient-local suppliers relations	d. Good	d. Good	d. Hierarchical	d. Good	d. Good	d. Good	d. Good

Organizational design for knowledge transfer:							
a. Coordination/ integration	a. Being attempted	a. Excellent	a. Being attempted	a. Excellent	a. Excellent	a. Being attempted	a. Being attempted
b. Concurrent processing	b. Being attempted	b. Excellent	b. Being attempted	b. Excellent	b. Excellent	b. Being attempted	b. Good
Opportunities to practice:							
a. Training and practice	a. Yes	a. Yes	a. Yes	a. Yes	a. Yes	a. Yes	a. Yes
b. Associated knowledge	b. To some extent	b. Yes	b. To some extent	b. To some extent	b. Yes	b. Yes	b. To some extent
Mastery: evidence of knowledge transfer	*Objective knowledge:* Yes – new products introduced; limited local design capabilities	*Objective knowledge:* Yes – new products introduced; local design capabilities limited	*Objective knowledge:* To some extent – new products introduced; local design capabilities limited	*Objective knowledge:* Yes – new motorcycles introduced; R&D capabilities lagging	*Objective knowledge:* Yes – manufacturing capability developed; only CKD assembly	*Objective knowledge:* Yes; production & R&D capabilities improved; new products introduced	*Objective knowledge:* Yes; production and R&D capabilities improved; new products introduced
	Subjective knowledge: Problem-solving and market orientation	*Subjective knowledge:* Problem-solving and market orientation	*Subjective knowledge:* Problem-solving and market orientation	*Subjective knowledge:* Problem-solving and market orientation adopted; 'world	*Subjective knowledge:* Subsidiary of 'world class' organization learning about Indian market	*Subjective knowledge:* Problem-solving and market orientation adopted;	*Subjective knowledge:* Problem-solving and market orientation adopted;

Table 4.2 (continued)

	Firm 1	Firm 2	Firm 3	Firm 4	Firm 5	Firm 6	Firm 7
	adopted; 'world class' identity	adopted; 'world class' identity	adopted; 'world class' identity	class' brand and high quality		'world class' identity	'world class' identity
Stage II Institutionalizing learning:	Technology collaborator acquired equity stake in the company	Joint venture partner acquired a majority stake; now a unit of a foreign MNC	Expanded product line and continued adaptation	Continued joint venture; introduced new products; continued adaptation	Continued to add new products	Continued adaptation & local innovation; introduced indigenously designed LCVs	Formed a 50–50 joint venture with Japanese partner

Note: * These assessments are derived from interview data based on questions about each of these categories.

Thus, an intriguing conclusion is that specific transfer mechanisms (i.e., joint venture, technical collaboration, or direct import) are less relevant than the transfer process and the creation of a community in achieving objective and subjective knowledge transfer. The presence of a social community is critical in providing access to tacit knowledge and relevant codified knowledge and artifacts. Learning in Bearings became possible when impediments to knowledge were eliminated in 1994, allowing the formation of a stable community of practice. Learning in Bearings was also possible when local expertise in certain domains (as in the heat treatment area) exceeded the level of expertise available in the parent firm. Here, US experts' confidence in local engineers' capabilities eased information exchange between engineers in the United States and India and facilitated experimentation and problem-solving at the recipient site. Particularly in the initial stages of transfer, knowledge is 'situated' in the community of practice because it is in the context of practice that questions emerge, solutions are improvised, and problems are solved. The subjective knowledge of identity and membership arise from interactions and negotiations with foreign suppliers that clarify and locate the position of the recipient in the international network.

Objective or substantive knowledge about new manufacturing processes, testing, benchmarking, and other best practices, and organizational design also resulted from practice. Its growth was initiated by the community creation process in the context of the project, but continued autonomously even after the project was completed and the community disbanded. An interesting consequence of adopting new technology is that all recipient organizations were compelled to assess internal capabilities, thus facilitating further infusions of knowledge into the firm and transfer to local suppliers.

In addition, an emphasis on testing processes to shorten cycle time, lower costs, and improve quality resulted in the creation of artifacts such as new documents and design adaptations to externalize tacit knowledge. Thus, problem-solving in practice generated more objective knowledge (new theories of organization; developments in technology) and new problems to be solved (for example, how to increase exports and improve infrastructure to generate domestic demand). Recognition of the benefits of specialization led to unbundling of different types of expertise such as expertise in design, manufacturing, and management and the perception of knowledge as a commodity. Consequently, in new transfer projects, as in the case of the cold rolling mill project in Steelworks, firms searched for specialized knowledge from various sources for project implementation. Thus, while the creation of community was critical in initiating the transfer, in later stages, recipient firms engaged in innovation autonomously because they had learned the analytical capabilities required to operate in the world of artifacts and objective knowledge through practice and participation.

Learning through practice in a cross-border community also resulted in initiating internationalization by helping recipient firms gain an understanding of what being global entails and aligning them with other global firms. While a large sample study is required to corroborate this conclusion, it is consistent with Cantwell's (1989) notion that foreign investment in locations with a capacity for innovation can generate virtuous cycles and spawn new multinationals. This framework of knowledge transfer suggests that the extension of firm boundaries depends both on relationships and community creation across firms and on internal capabilities. With continuing involvement in practice, the relevance of a particular community may decline as technology recipients develop capabilities and engage in refining their own technological trajectories. They then attempt to access knowledge in other communities. Hence, developing organizational knowledge of how to collaborate is particularly important for transnational firms.

Finally, practice is central to building communities, developing objective knowledge, and learning collaboration skills. It is the keystone in structures that facilitate cross-border and other knowledge transfers and knowledge creation.

NOTES

1. CKD = completely knocked down.
2. CIF = cost, insurance, and freight.
3. Questions were adapted from studies by Georgopoulos and Mann (1962) and Van de Ven and Ferry (1980) to assess communication.
4. I used questions adapted from studies by Georgopoulos and Mann (1962), Clark and Fujimoto (1991), and Tyre and Hauptman (1992) to assess coordination.
5 Computer-aided process planning and computer-aided manufacturing.
6. ISO (International Organization for Standardization).
7. Calculated at the 1996 exchange rate of INR35 = US$1.00.
8. At an exchange rate of US$1.00 = INR34.
9. These comparisons were obtained from questionnaire data from the three focal firms in this chapter (see Surie, 1996 for details).
10. As above, these comparisons are derived from the questionnaire data.
11. The last time a new blast furnace was installed was in the 1950s; it was then the largest in Asia. Since then, there have been many advances in blast furnace and iron-making technology.
12. mtpa = million tonnes per annum.
13. tHM = kilograms per tonne of hot metal.
14. With the exception of one firm that had recently experienced a labor strike.
15. The need for administrative and social systems to match technical systems is a common theme in sociological literature.

5. Accelerating innovation in manufacturing – architecting complexity: Stage III

In this chapter, I focus on adaptive responses that helped shape the development of complex organizational systems[1] in the three manufacturing firms described in the previous chapter (Steelworks, Bearings, and Earthmovers) as they sought to globalize in the new millennium (post-2000). Increasing global competition required continuous adaptation by organizational restructuring and value chain reconfiguration to respond to competitive challenges, uncertainty, and complexity.[2] Improved manufacturing capabilities enabled these firms to gain autonomy and enhance their position within their parent network, thereby leading them to assume greater responsibility, expand internationally, and develop a global perspective. Although apprenticeship was the chief mode of learning in the earlier stages, during this stage the emphasis was on transforming the organization to support larger-scale operations while fostering local innovation. This transformation entailed shifting to an industrialized system of knowledge production by adopting a modular organization and relying more on codified knowledge to accelerate innovation. It also required aligning the financial, strategic, and operational goals of the organization to compete in a global market.

New organizational capabilities developed in earlier stages by adopting lean manufacturing systems and using advanced information and communication technologies set the stage for coherent global expansion and integration. By aligning overall strategy with every functional area, firms were able to achieve control and manage internal reorganization to match strategic aspirations. Past success in achieving goals of acquiring and assimilating new technology led executives to view themselves as participants in a global economy. Leaders of these firms concentrated on making the organization more efficient and acquiring capabilities to position themselves as the lowest-cost knowledge-intensive and technologically advanced firms. Consequently, the discourse adopted emphasized themes from the popular business press and management literature of the 1990s.[3] The following sections examine three principal mechanisms that aided the transition to a more complex system:

1. the adoption of modular and flexible organizational structures, resulting in the emergence of *heteromorphic* organizational form, a hallmark of which is the use of multiple organizational designs concurrently and over time to accommodate increasing scale, complexity, and size;
2. the use of information technology for all activities to accelerate the pace of knowledge decomposition and enable the recombination of tacit knowledge to create new knowledge (new rules, categories, and models) and products (knowledge components);
3. the pursuit of a *real options heuristics* (ROH) strategy to achieve goals and aspirations.

These steps, together with greater awareness of the global arena, precipitated a further shift in cognitive orientation: the perception of the firm as a bundle of assets and capabilities and a realization of the potential of strategic and financial management methodologies. Consequently, such methods were deployed to leverage the firms' assets and accelerate global expansion. A detailed analysis of each firm's evolution to complexity using the above framework follows.

STEELWORKS: IN QUEST OF GLOBAL EXPANSION

Since the early 1990s, Steelworks experienced continuing pressure to become globally competitive, particularly since investors and analysts had shunned the steel industry in favor of e-commerce and software where returns were higher. Therefore, re-engineering the company to meet financial targets was critical to establish credibility with the investor community. As in earlier stages, business process re-engineering accompanied the implementation of new technologies in turnkey projects as the company moved from producing commodity steel to making value-added products required for the automobile and appliance industries.

Transforming Steelworks into a leading global manufacturer involved re-evaluating assumptions underlying corporate strategy. The strategic planning function moved from a five-year cycle to an annual cycle, and finally, became a continuous process. Thus, strategy evolved from an isolated function into an integrated part of the whole business. According to the chief of strategic planning:

> In 1997, strategy development was basically led by engineering planning and financial planning . . . by 1998 it evolved to include fairly well-articulated long-term strategies and guidelines for making an annual business plan. Yet it was very much a stock-driven activity at this time . . . it used to occur at a very specific point in time in the year. By 2000 it had become a calendar-driven thing . . . and

> by 2001, strategy development started resembling more of a continuous structure rather than something that started only in October, recognizing that strategy is indeed a proactive response to stimuli whether these are external or internal.

Early efforts to develop strategy emphasized technology. However, now top management focused on developing a corporate perspective, and building organizational capabilities for growth and competitive strategy. This transformation was not merely rhetorical and pervaded all divisions of the organization. Growth strategies were reviewed in quarterly workshops using McKinsey's three horizon framework; similarly, financial models were used to project short-term cash flows and price forecasting capabilities were enhanced. Strategic thinking and the strategy process were institutionalized throughout the company. All profit and cost centers, functional divisions, and business units were exposed to these methodologies, to engage every member of the organization. An interviewee pointed out that a function of strategy was to serve as a communication device: 'The strategy needs to be created, and people need to be involved in the process of creating that strategy. People need to know the strategy so that these strategies are aligned to the goals of the company'.

Adopting Modular Organization

Integrating strategy across the organization called for organizational structure to match the evolution of the strategy process. Reorganization at Steelworks was accomplished in four phases. In the first phase (1994–96), the focus was on attaining desired volumes in the core business. In this phase, the structure remained a functional hierarchy. In the second phase (1997–99), management emphasized cost competitiveness and a market focus, with the attention centered on net realization and costs. Reorganization was begun through changes in sub-units that focused on new measures for productivity, customer satisfaction, and efficient manufacturing (as delineated in the previous chapter). Organizational reconfiguration continued in the third phase, which focused on 'earning the right to grow' (2000–02), while the fourth phase (2003 onwards) could be characterized as a growth phase with an emphasis on financial measures such as net present value (NPV), return on invested capital (ROIC), internal rate of return (IRR) and economic value-added (EVA).

Post-2000 efforts were focused on adopting a modular hierarchical organizational design to execute strategic plans effectively. Models commonly used in industrialized countries were adopted for strategic planning. Plans were also constructed hierarchically and consisted of a five-year market plan at the apex. Translating such plans into cash flows over a five-to-ten-year time horizon involved identifying growth opportunities. To do so, analytical tools such as portfolio choice models were used to analyze opportunities and make decisions about which businesses the organization should pursue and which ones to abandon. Thus, a real options heuristics (ROH) strategy seemed to have emerged in which opportunity-seeking and experimentation were driven by the need to obtain more information both to reduce uncertainty and to achieve new aspirations. Implementing this strategy required reconfiguring the organization into modules or units to enable management to align decisions with financial objectives.

Modularity, a strategy for designing complex technical systems and processes efficiently, is a design rule that facilitates concurrent information processing (Baldwin and Clark, 2000).[4] Components are joined through interfaces in technical systems. Similarly, organizational systems consist of sub-units or modules that operate independently connected by interfaces. These permit high levels of intra-unit coordination and enable the organization to function as an integrated whole. Modularity is achieved by partitioning information into visible design rules and hidden design parameters, and is efficient only when the partition is unambiguous and complete. Adopting a flexible, modular structure entailed partitioning the enterprise into independent modules that were linked together with standard interfaces.

Modularity at Steelworks was achieved in stages. In 1997, Steelworks was a functional organization with several divisions including marketing, product, raw materials divisions, and a three-layered management structure consisting of senior, middle, and junior management. In 1998–99, although the functional organization remained with many divisions, a fourth layer of structure was introduced. In 2000, the marketing and sales organizational structure was reorganized to include focused customer account managers leading the sales division and the customer interface. In 2001–02, the functional organization was abandoned and profit and cost centers were established, reducing hierarchy and flattening the organization. Modules such as strategic business units (SBUs), cost and profit centers were created and linked through rules aligned with the overall organizational strategy to facilitate the manipulation of individual sub-units. These rules consisted of financial parameters such as EVA (economic value added) and NPV (not present value) or quality standards such as ISO to help evaluate projects across the organization. Nevertheless, administrative structures and functional areas requiring a high degree of special expertise (such as finance) were retained.

Facilitating Knowledge Decomposition

Partitioning organizational hierarchy into a modular decomposable system requires creating stable sub-systems and sub-assemblies to facilitate knowledge decomposition[5] and innovation. In problem-solving tasks, partial results that represent recognizable progress toward the goal play the role of stable sub-assemblies and can be considered as assets or components in the whole task.[6] Systems that are composed of sub-assemblies are capable of evolving faster (Simon, 1956). Viewing the accomplishment of strategic goals as a problem-solving task for the overall system, sub-assemblies consist of units that may be deployed in different combinations to fulfill organizational goals. The use of new information technologies facilitates knowledge codification and decomposition, and enables task-partitioning and the creation of sub-assemblies and components (Cohendet and Steinmueller, 2000). By incorporating information technology into all activities it was possible to distribute tasks and leverage knowledge generated from different units and sub-assemblies (such as multifunctional project teams and task forces). Such units could also be assembled and disbanded speedily as required.

In addition, organization-wide adoption of new information technologies eased coordination and communication by facilitating sub-system standardization and integration. In Steelworks, coordination between sub-assemblies was fostered by establishing norms and rules[7] to regulate behavior. Such rules were disseminated at each level of the organization via communication emphasizing the vision, mission, strategic direction, and goals. Likewise, coordination was established by disseminating concepts, methods, and techniques (such as organized cost reduction, value engineering, benchmarking, and the product portfolio matrix) to standardize the output of sub-units. New programs were developed to disseminate and apply these rules systematically for domain-specific tasks. An example is the 'Total Operating Performance' (TOP) program for manufacturing and marketing. Similarly, by introducing a 'Performance Ethic Program' to identify and match employee strengths with job demands using NPV analysis, risk analysis, and financial tools, top management aimed to build a new culture that emphasized value creation and a financial orientation.

The process of modularizing the organization continued by emphasizing knowledge management and by continuous attempts to codify tacit knowledge embedded in the entire value chain of activities. Rendering knowledge explicit also resulted in developing new controls, parameters, and rules.[8] For example, embedding codified knowledge in equipment reduces the effort required to control sub-systems to an examination of pre-specified parameters. Thus, codification allowed new parameters to emerge and facilitated both intra-unit and inter-unit coordination. The use of new technologies to track and codify the knowledge and competence embedded in each organizational unit and expert also facilitated the creation of knowledge components that could be treated as exchangeable assets or real options.[9]

Similarly, in the areas of strategy and organization, new parameters for gauging success in achieving organizational goals were developed through mastery over change processes acquired through workshops and special projects conducted with the help of consultants like McKinsey & Company, Booz Allen Hamilton, and Arthur D. Little.[10] In addition, top management also sought experts with specialized technological knowledge in steelmaking to learn about implementing large-scale projects for value-added products. Experts included leading steel manufacturers like Thyssen of Germany, Hoogovens of Belgium, and Nippon Steel of Japan. Technology and strategy workshops were conducted for organizational units at corporate, profit center, and functional levels. This led to the adoption of new parameters to link sub-assemblies, thereby facilitating the manipulation of the overall organizational system.

The use of new communication technologies to decompose knowledge by codifying various processes exposed redundancies commonly found in complex hierarchical systems,[11] which are usually composed of a few different kinds of sub-systems in various combinations and arrangements. Such knowledge decomposition helped to standardize knowledge replication by diffusing various methodologies and assumptions. In addition, it also helped to accelerate innovation by speeding the recombination of knowledge generated from different sub-units, sub-assemblies, and projects.[12,13] Examples include the diffusion of methodologies used for strategic and financial analysis. As before, financial measures such as EVA, IRA, and NPV, and tools like scenario planning were used to control and standardize sub-units and make predictions, thus permitting intra-organizational comparisons. In finance, Steelworks drew on consultants like Stern Stewart, McKinsey, and A.T. Kearney. When the managing director established the goal of becoming 'EVA positive', all project analyses focused on this measure. New projects were evaluated based on their likely contribution to shareholder value, probability of success, and time to profitability. Standardizing financial criteria for resource allocation decisions and for evaluating all activities allowed comparisons across disparate projects. Consequently, organizational sub-units lacking current or future potential were abandoned; in contrast, new projects including collaborative projects and strategic alliances were regularly assessed to determine ongoing resource allocation. Besides emphasizing specialization and new modes of organization, a new finance-oriented culture was disseminated based on the assumption that the raison d'être of the firm was to create shareholder and market value. This new philosophy was articulated as the 'Steelworks Business Excellence Model' and implemented to deliver high performance and quality.

Finally, in an uncertain, competitive environment where speed was critical, management abandoned the policy of attempting all projects internally. By using these methods and financial parameters, activities could be compared from the perspective of what each contributed to the bottom line. Thus, innovation was accelerated by standardizing and industrializing the knowledge production process. Standardization and comparability also led to specialization. The emphasis on adding value resulted in abandoning peripheral activities that were part of the organization's heritage, including those that stemmed from the nation-building stance adopted at founding. Consequently, a department was established to outsource municipal activities such as town and water management.[14] However, in some instances such as hospitals and health services, financial considerations, while important, did not drive divestitures.

Similarly, port operations and international business were divested to another part of the conglomerate. This shift in orientation led to the insight that instead of just outsourcing technology, it was possible to manage technology actively by evaluating and changing it to meet current requirements. However, as an integrated steel plant, many upstream businesses such as iron ore mining, which provided the lowest cost inputs in the steel industry, were retained. A manager noted that because the industry was process-oriented, outsourcing all processes was difficult.

In addition to outsourcing peripheral operations there was an emphasis on forming alliances where partner capabilities could be leveraged for learning. Several alliances had been established, including two recent joint ventures. The first was to produce a critical desulphurizing compound (since Indian iron ore contained high levels of sulphur). The second, formed with an Italian firm, was to handle import operations to take advantage of both partners' capabilities in different Indian locations. Thus, both alliance and hierarchical organization forms were in use concurrently.

Real Options Heuristics

Management also began to apply portfolio concepts and use a real options heuristics strategy to evaluate businesses. Thus, maturing businesses facing reduced returns because of low entry barriers (such as the refractory business and sponge iron) could be exited, while attending to developing new capabilities through exploration. Management divested a cement plant to raise cash and fund new businesses and activities likely to improve market position. Through brainstorming across the organization, initiatives emerged and evolved into business units. One example was a profitable business in a mineral-based industry using access to a key resource, Indian reserves of titanium dioxide (among the best in the world), to manufacture pigments for paints.

The notion of using information technology and real options heuristics for strategic decisions also pervaded the R&D organization. This evolved over five generations from a peripheral activity done in isolation to one that was integral to the organization. The knowledge embedded in the R&D organization was also used for strategic decisions about how to select and optimize current projects and which technologies to explore in areas with the potential to produce breakthrough innovations. In the research area, there was increased emphasis on using mathematical modeling and simulation to understand steelmaking processes and evaluate the properties of steel. Simulation methods were also extended to associated departments, leading to wider use of this methodology. Physical models were built for processes that were too complex for mathematical simulation. Such experiments had led to the development of an off-line simulator to evaluate the properties of steel; an on-line simulator to predict the properties of steel while being rolled was also being developed. Expertise developed over time in specialized areas of steelmaking such as beneficiation was provided as a fee-based service to other companies. In addition, R&D obtained data and suggestions from other areas (such as the mines) to update and create new systems. Although 75 percent of the R&D work was focused on current projects, 25 percent was directed towards exploration (March 1991). Exploratory projects focused on areas such as altering the properties (e.g., strength) of steel. Along with R&D advances, intellectual property management became a major focus. Although Steelworks had begun to patent knowledge (with 23 patents obtained in one year after a total of only 73 in the previous 90 years), much work lay ahead. Although the R&D head emphasized that: 'Being innovative and having patents, they are not one and the same thing', yet patenting was being pursued more vigorously. This was as yet a defensive measure, largely because of the threat of being sued for infringement of intellectual property rights.

Access to new knowledge was obtained by pursuing collaborations via special projects at leading universities in Germany, the United States, and the United Kingdom, particularly with researchers at centers reputed for their expertise in steelmaking. Many researchers were retained through contractual arrangements as consultants to train internal experts at Steelworks. Realizing the shortcomings of internal R&D, the organization also sought external experts and membership in industry associations to acquire knowledge and build capabilities. These changes were accompanied by a keen awareness of the importance of managing and producing knowledge in the company.

Though not yet a management problem, a challenge from focusing continually on new opportunities was to find resources to manage multiple investment opportunities, since 'the clubbing of investments' could lead to 'a cash flow mismatch situation'. In addition, managers faced challenges arising from changing industry structure, tariffs and subsidies that dictated demand in world markets, and the company's cost structure. Finally, the need to manage the organization's cost structure was another source of concern. For example, the wage cost in Steelworks was approximately 18 percent of turnover compared with the world average of approximately 10 percent; in contrast, a steel company in China had a cost structure of 6 percent of sales, providing it with a clear advantage. The company also faced other challenges such as retaining talent in a competitive environment, deteriorating ethical standards, and building relationships with the new state government to retain competitive advantage. A strategy chief explained: 'We are a first world steel plant operating in a third world environment'. Nevertheless, managers exhibited a sense of confidence and control, having gained mastery over tools that provided new ways to manage uncertainty.

By the end of 2006, the company had become a supplier of value-added steel to all international white goods companies. A mergers and acquisitions (M&A) manager noted that Steelworks was one of the only two integrated steel producers with a presence throughout the value chain, making them less vulnerable to business cycles. Aspirations continued to rise with a new target set for becoming a larger global player with a presence in multiple global locations and production of over 30 million tonnes per annum (mtpa) by 2015. New rules based on past learning evolved. These included achieving expansion through acquisitions or joint ventures overseas in the case of greenfield projects[15] to accelerate the process and using a distributed value chain across the globe to optimize costs and gain flexibility.[16] At this point, the strategy shifted to establishing primary steel production in resource-rich countries to take advantage of ownership of strategic raw materials (e.g., ore, coal, and gas).[17] Thus, finishing facilities would be located closer to the customer in growing markets such as Southeast Asia, Vietnam, and China. Expansion would also be facilitated by an increased focus on logistics and branding. Examples of this strategy include a joint venture with an Australian company to manufacture steel for pre-engineered buildings. The expansion-by-acquisition strategy culminated in the acquisition of a UK-based steel company in early 2007, catapulting the company into the ranks of the world's top five steel producers.

Top management noted that this expansion would surface new challenges in the areas of human resource management, finance, relationship management with stakeholders, and managing cultural differences. Most important was the need to create a flexible organization when operating from multiple locations in different countries while retaining the values of a strong community in a single location.[18]

BEARINGS: CONSOLIDATING THE TRANSNATIONAL NETWORK

The acquisition of the Indian joint venture by the US partner helped raise awareness of the potential of outlying subsidiaries in emerging markets and helped the US parent evolve from a company with international operations to a transnational organization. Moves made by MNCs in the manufacturing sector to establish themselves in emerging markets hastened the evolution towards tighter global integration. To make Bearings more customer-oriented, the US parent reorganized around customer segments in 2000. This helped to transform a multidomestic company into a more integrated global organization. Yet, subsidiaries in India, China, and Eastern Europe continued to function independently under the aegis of 'emerging markets' and were not merged with the main businesses. However, from about 2003, the Indian subsidiary became a part of the global industrial business segment after a fresh reorganization. Using the framework outlined above, I document the transition to increasing globalization and organizational complexity.

Adopting Modular Organization

Modularity as an organizational approach had been attempted earlier while establishing the Indian manufacturing plant. However, this was aborted initially, partly because of distance and communication and coordination difficulties, and also because multiskilling and training required for manufacturing flexibility were too expensive and time-consuming. Cross-national coordination became easier after a 1996 reorganization in India towards a more integrated system. Subsequently, the manufacturing organization became more deeply embedded in the global organization because of its altered position as a subsidiary. This was reinforced by technological changes paving the way for decomposition of the value chain of activities and major reorganization at headquarters in 2000. As a top executive noted:

> It is now one of the 'Bearings' plants and can supply anywhere in the world. . . . Today everything is on the international network of human resources and . . . it has of course automated a lot in the non-manufacturing areas as well. . . . When I started there were 13 people including one director. Over the course of the last seven years, the department has hardly six or seven people now.

In addition, a matrix structure was used to integrate organizational modules across the global value chain. Executives at the Indian manufacturing sub-unit executives reported to both the subsidiary head and to corporate headquarters in the United States. However, a matrix structure was not adopted uniformly; some executives had a direct reporting relationship to one group. Other integration approaches included sending employees on short-term assignments from India to other parts of the world to stimulate knowledge flows and leverage the knowledge gained through replication in new locations. As a top executive noted:

> Because of our becoming a part of this international organization, there is a lot of focus on talent management . . . that area is much more systematic and structured than what we were doing earlier . . . one of our associates has been identified to go as a manager to South Africa. The other thing in terms of talent management is that there is a lot of opportunity for people to work on international assignments, short-term assignments, rather than long-term deputation there.

An example was a project that involved equipment systems across the globe. Employees from the Indian manufacturing unit were also actively involved in establishing the Chinese plant.

Despite the adoption of modular organization, according to a manager of operations and logistics, integration with the global organization required adopting new approval processes that slowed decision-making. Approval of vendors by the country head and global headquarters was time-consuming and caused delays. However, interventions by managers in these locations served to reduce errors by providing additional input.

In contrast, another manager noted that in the early 1990s, when the company was a joint venture, innovation was thwarted because engineers and technicians in India were taught 'know-how' but not 'know-why' and had to operate within tight boundaries. After becoming a subsidiary they were able to debate the best way to make something or try out other solutions.[19] Ironically, local managers often had more freedom when headquarters did not have a solution. Moreover, according to this manager, the perception of 'made in India' had changed. Also, experience in India over the past ten years made it easier for US-based managers to trust Indian employees regarding quality. Consequently, local employees were not using completely traditional methodology developed by Bearings.

The adoption of a matrix organization required some senior managers to attend quarterly meetings at headquarters in the United States, simplifying the task of transmitting new ideas (such as quality-related six sigma processes) and practices across the organization through a best-practice sharing system. Moreover, the rules of the game were made very clear – that the lowest-cost plants would be favored. Thus, local engineers were studying ways to reduce scrap and consumables and introduce process systemization. While the cost of human resources was not a major issue, the HR executive noted that as labor costs increased, this might become an area of focus, particularly since the Indian plant was more labor-intensive than those elsewhere. In addition, automating manufacturing processes would reduce handling and improve quality.

Integration was also induced through common systems used to gather data in all locations. Headquarters was provided with 'all the customer data – it is very transparent and they have access to any data they want'.

In addition, integration between the manufacturing plant and R&D in Bangalore was fostered by transferring individuals from the former to the latter. A manager from the manufacturing unit had recently been moved to Bangalore and made the head of global sourcing from India. Similarly, the strategic purchasing function, earlier managed by the manufacturing unit, was moved to the country headquarters in Bangalore. Thus, knowledge exchange between manufacturing and R&D was encouraged. Since India was now identified as a potentially important source of global supply, information technology, engineering and management skills, the Indian subsidiary played a more critical role.

Intensive training was critical to ensure that every level of the enterprise became capable of delivering high-quality output and that interfaces between different organizational modules functioned efficiently. Scientists and engineers from Bangalore were sent to the United States two to three times a year for training. The subsidiary head observed that: 'Many of the engineers spent four to six months initially going through intensive training in-class in the US, detailed engineering training in doing analysis, [learning] what they have to do, customer issues, what they have to do in the field'. He noted that the training required significant investment in time and cost in production and manufacturing areas, but only a few weeks in IT.

Facilitating Knowledge Decomposition

The headquarters reorganization and adoption of automation to rationalize global production and increase productivity helped partition activities and led to increasing specialization across the global value chain. The Indian plant allocated the industrial bearings segment while the Chinese plant became a supplier for the automotive segment. While an R&D unit was established in Bangalore, India, the company also decided to use this location as a base for global sourcing. These changes in context led to the evolution of new rules.[20] For example, restructuring the organization to incorporate the Indian subsidiary in the global value chain dictated shifting from a domestic to a global orientation and focusing on customer segments worldwide rather than solely by geographic location. Thus, by 2005, a new product line was introduced and manufactured by the Indian subsidiary for customers in the United States. In early 2007, managers noted that order fulfillment was centralized in the United States. Automation was adopted to standardize, tasks were standardized to the point where conformance to stringent quality standards was inevitable because: 'a single process [was used] to drive the system worldwide'. Part numbers were distributed across the world and some part numbers were sourced only from India. Decisions were based on a determination of lead times and economics of each location and plant: 'Each plant has its own capacity and number of machine hours available; each part number has a cycle time'.

These close interactions in the process of embedding a modular organization within the multinational network strengthened the relationship between Indian subsidiary managers and those at US headquarters, resulting in a more equal cross-border partnership.

Another consequence of the focus on quality at the lowest cost led to an abandonment of the strategy of sourcing everything internally, indicating further evolution and diffusion of market-oriented principles across the multinational network. Thus, 'quality at the lowest cost' could be viewed as a new market-oriented rule for sourcing supplies:

> It is one of the Bearings' plants. And it can supply anywhere in the world. Likewise, for our sales and marketing people, what we need to sell in India, they need not necessarily buy from us. They can go to Romania . . . we would have to compete with other units. So, it's a much broader span now.

The critical 'make or buy' decision was another important rule that was modified. It was altered to take account of new conditions such as the availability of external capacity and bottlenecks in the internal supply chain. A manager noted that manufacturing had become a bottleneck because of the need to supply worldwide, and supplies were insufficient. Hence, to keep customers satisfied, it became necessary to treat manufacturing as a vendor in the supply chain. In 2002, a decision was taken to limit in-house investments because of favorable external supply conditions. Consequently, 70 percent of turning operations, 5 percent of grinding operations, and 15 percent of the heat treatment operations were outsourced.

In addition, using information and communication technologies (ICTs) surfaced problem areas by making interfaces between departments more visible. As solutions were discovered and new parameters evolved, they were encoded into the system. A manager observed that their focus was to make the supply chain visible, eliminate inefficiencies at the boundaries of departments and increase accountability: 'Each department says this is not my problem . . . the supply chain is hidden in three departments'. Additionally, the intent was to develop 'robust systems that are not man-dependent' so that customer orders were visible to every one with access to the system.

Moreover, interactions with headquarters and application of statistical methodologies associated with standardizing processes led to an awareness of the need for empirical data and scientific analysis for problem-solving and decision-making. Adoption of IT-based systems helped to reinforce the notion that decisions should be based on data rather than intuition. For example, if a customer ordered 300 pieces of a certain bearing, each input was changed to compensate for suppliers' inability to provide timely delivery, thus increasing the order size and resulting in 'a lot of dead inventory'. However, the main impediment to efficiency was not related to the supply chain but a result of the attitudes and behaviors of individuals.[21] Changing such behavior would require the application of 'soft strategies' since people were used to 'estimating' rather than counting or obtaining exact data. The manager noted that: 'We have a lot of systems . . . very few people are actually using the data and systems. [In the] initial [stages], the data were not good . . . it is up to the user to go to the systems department to push to change it . . . adoption of the system is not automatic'.

A central objective was to improve the formulation and implementation of plans and optimize resources via quantification and statistical analysis. The manager noted that it was necessary to: 'document each sub-process with the mean and standard deviation to create a mathematical model and reduce either the mean or the standard deviation'.

Another specialized unit, the R&D facility, had been set up in Bangalore recently to take advantage of low-cost engineering skills available in India. One of the company's global R&D centers, it focused on three broad areas – product technology, manufacturing technology, and information technology. Along with the other global R&D centers, the product technology group focused on global customer groups in the United States and Europe and product performance, which involved the quality and audit functions. The manufacturing technology group focused on grinding/finishing, automation controls, tool design, and engineering detail design, and used electrical and software engineers to support working with suppliers. In IT, the subsidiary was integrated tightly with global headquarters; the top management team planned to locate at least 40 percent of the firm's IT capacity in Bangalore. By 2007 this IT center was the global operations and corporate shared services back office with accounts receivable, payroll, and a 150-seat call center.

Besides being a low-cost location, India also offered a skills advantage due to the presence of an 'analytical and numerically savvy workforce'. The head of the R&D center in Bangalore noted that the goal was to determine which core areas to focus on so that: 'we don't keep adding people just for low cost . . . we get good at what we do. Like on the customer engineering side, we focus on a specific product of a certain size range; India is going to be the global center of excellence for advanced analysis modeling, because we are good at that'.

The new facility would also focus on knowledge management and non-traditional products. Knowledge management covered specific types of marketing, engineering, or R&D expertise, while non-traditional products were included in India because of the availability of new engineers from different industries – 'they come up with new perspectives, new ideas, as opposed to somebody who has worked in our headquarters for 20 or 30 years'. The subsidiary head noted that specialization was critical to remaining competitive over the long term:

> Any company that comes here needs to understand what their true competencies are and quickly decide in six months or a year, after looking at the talent of the people, what are the key areas they can really focus on. And then focus only on those until they get good at it. Because in any company the tendency is to do 50 different things . . . until finally they get nothing.

Real Options Heuristics

The Indian subsidiary of the MNC was integrated into the global organization and specialized in manufacturing specific components, some R&D and new product development, back office operations, and IT services. Consequently, the MNC followed a limited ROH strategy. It remained focused on optimizing global operations through sourcing from India.[22] Within India the adoption of an ROH strategy was limited to outsourcing to local suppliers and expanding local operations by replicating facilities. Thus, in 2007, a new manufacturing plant was to be established in Chennai.

The local environment had also evolved and economies of scale had changed as the auto-component industry was experiencing a CAGR[23] of 12–15 percent as a result of US automobile companies' attempts to slow their decline by sourcing from lower-cost locations. Consequently, new specialized firms had emerged, greatly enhancing logistics[24] and making it possible for the firm to shift to a model in which more work could be outsourced to local suppliers. This was particularly relevant since Bearing's Indian operations focused on optimizing production and rationalizing activities via outsourcing. Although growth by acquisition was a possibility because of aggressive growth targets, such decisions would be made by top management in the United States.

Both the R&D and HR heads noted the challenge of nurturing and retaining talented employees. The HR head focused on changing the culture of the organization and motivating employees in a local unit of an integrated global organization with customers located in different countries.

Hence, the challenge was to motivate employees to be the best unit of a larger organization. In contrast, the R&D head noted that:

> I looked for people with good educational background, from IITs [Indian Institutes of Technology] and regional colleges, and who have strong family ties . . . in fact many of them studied in the US and came back to India . . . but it is not a huge pool. . . . People like that want to be part of India's transformation . . . which would really benefit the country, the company, and the individual.

Attracting and retaining talent continued to be the chief challenge in 2007, particularly in the main manufacturing plant, which young engineers did not perceive as a choice destination. This difficulty was compounded by rising salaries[25] fuelled by the growth of the auto-component and automobile industries as operations of many other US firms shifted to developing countries.

EARTHMOVERS: BUILDING AN AUTONOMOUS ENTERPRISE

Evolving from the construction equipment division of the automotive unit of a conglomerate, Earthmovers gained autonomy as an independent firm in 2000. Simultaneously, its collaboration with its Japanese technology supplier developed into a joint venture, strengthening partner ties and making the firm more aware of the need to focus on quality and customers. Consequently, the pace of improvements accelerated not only in manufacturing, but also in R&D and design. Growth and the evolution of technology dictated corresponding changes in organization. A new plant was established in Southern India for indigenously designed smaller excavators. Changes in organization design are outlined below.

Adopting Modular Organization

Originally a division of a truck and automotive company, Earthmovers was spun off as an autonomous subsidiary when it began to contribute about 10 percent of the company's revenue. The Japanese partner took an initial equity stake of 20 percent in the independent entity, and increased it to 40 percent in December 2005. In the mid-1990s, management began to adopt a modular approach to organization to enhance problem-solving, mirroring the manufacturing technology adopted from the Japanese partner.

The organization was structured to take advantage of information generated in all areas. The Japanese partner was very cooperative and was more involved in manufacturing, quality, design, and service support than during the earlier phases. Three Japanese expatriates were stationed in the main manufacturing plant to provide assistance on a day-to-day basis. The Japanese head of manufacturing quality was also on the board of the joint venture. Consequently, organizational design, practices, and methodologies used by the partner firm were adapted for local use by the Indian partner. More attention was paid to quality with the full-time presence of a Japanese technical expert to oversee quality assurance.

Since the establishment of the joint venture, the number of design engineers had increased along with design productivity and the rate of new product introductions. Instead of only one product per year, the company had begun to introduce two or three models per year. These included both indigenously designed products and those made collaboratively with alliance partners.

A new plant was established in Bangalore. The new unit also had a design office with eight R&D people who focused on design-related activities associated with the four models assembled at the new plant. Many products were not radically new, but were improved versions of earlier models no longer under license. Also, components were improved and used across different products. In addition, products were renamed to build the Indian conglomerate's brand.

Facilitating Knowledge Decomposition

As in other firms, the entire organization had adopted information technology extensively to accelerate innovation, design, and new product development. According to a senior executive, the new equipment and technology supplied by the joint venture partner was 'very IT-dependent', facilitating problem-solving by allowing problems to be decomposed into sub-problems. As a result, new knowledge components and products could be created by recombining elements and building blocks. This continuous process of knowledge decomposition and recombination enabled the knowledge of the firm to evolve by facilitating the discovery of new building blocks and the emergence of new rules for action.

> The new equipment has a lot of in-built technologies in the form of IT-based systems . . . whereby you can download or data log a lot of information which you normally require, in the form of . . . how many cycles of scoops the machine has brought out, how many times it has swung, how many kilometers of distance it has traveled during the day . . . all the data can be downloaded. One more step [which is not useful in India at the moment] is that if the machine is working in [x], one can see the operation of the machine on a PC as you can hook on that

> machine with your PC through the satellite connectivity. The machine has the capability to store data for 100 hours, which is roughly about three months' data . . . So it helps to really monitor the equipment.

Similarly, computer-aided design (CAD) had been widely adopted earlier along with other new methods such as the finite element method (FEM) to improve design accuracy. FEM is a method of solving problems by decomposing them into components and approximating the global solution. It is widely used for fluid mechanics, heat transfer problems, and for calculating stresses in complicated structures. Thus, it provides the ability to predict whether a particular part would or would not fail under stress, reducing the number of testing hours and iterations. More people had been trained in FEM analysis and the company was close to achieving the target of completing 60 FEM analyses in 2003. Each analysis was complex, the most difficult part being the fixing of boundary conditions. Other systems used by advanced companies included field data acquisition systems. In addition, they were considering establishing a testing laboratory within the next two to three years since testing was not being done as yet.

By obtaining detailed data on component prototypes and the conditions of products in use, the organization simplified new product development. Moreover, new product development capability was enhanced: the number of new products for different purposes arising from variability in the context of use increased. Indigenously designed products such as a 10-tonne wheel-loader and a new motor grader (designed six months earlier) were launched successfully. Other new models included incrementally improved variants of older models designed and manufactured under technical collaboration.

As a result of collaboration with the joint venture partner and greater familiarity with global players, top management revised aspirations upwards:

> The plan for this company is to become an INR10,000 crore (INR100 billion; approximately US$2.5 billion @ US$1 = INR40.38) company by 2011–12. . . . [We are] not only looking at the present business streams that we have but are also looking at other different products to become a full line company. Today we are only considered an excavator company. . . . We now want to add road products. We are looking at rubber-tyred equipment and trying to become one of the formidable companies in backhoe loaders, wheel-loaders, rear dumpers for mining use.

The use of information technology to partition activities across the firm also led to the emergence of a financial orientation and emphasis on logistics. Hence, the desired goal would be obtained through both organic growth and the addition of new business streams as above: 'If there are certain organizations or certain units which would synergize with our operations, we would not mind looking at them. . . . It will be a combination of both [organic and inorganic growth strategies]'.

Thus, revision of rules for processing information and making decisions included examining whether financial, market share, or customer service objectives had been achieved, and comparisons with competitors:

> We enjoy 55 percent market share and continue to be the leaders in the excavator market. . . . We have one of the best networks to support the equipment with something like 65 outlets all over the country. So we are virtually not more than 150 km or in a radius of 150–200 km from any customer location. . . . So that gives us the greatest edge from the point of view of holding on to our leadership position. And . . .
>
> I think the key differentiator [in meeting the challenge to sustain leadership] would be to ensure that we look after our customer well. Ultimately he is the king . . . any good product if it is not handled or supported properly will never be successful.

Executives had uncovered gaps that could impede achievement of aspirations by benchmarking performance against their Japanese partner. For example, in the design area, fundamental design work was still not done; much of the R&D activity was related to adapting products for the Indian market. While aware of the need for patenting, emphasis on patenting was lacking for historical reasons as the Indian parent company's engineering research center had been responsible for patenting. Evaluating performance in other areas, an executive noted: 'While on the one hand manufacturing is a great strength, we need to improve our manufacturing from the point of view of raising the standards and level of output and delivery, attention to customers, attention to the finer points'. Also, he observed that despite the presence of a good vendor base, they had not been able to achieve just-in-time inventory and were working on improving logistics.

Real Options Heuristics

A real options heuristics strategy was used primarily to build capabilities and gain leverage in the Indian market through association with a foreign firm to build the Indian brand. Earthmovers used multiple alliances to build a robust knowledge base that would allow it to cater to the rising need for increasingly specialized equipment. Thus, in addition to alliance partners from Japan, the United States, and Canada, late entry necessitated an alliance with the number two manufacturer in Spain for vibratory compactors.

The R&D head noted that competing with large firms was difficult because of its small size; its turnover was INR500 million (US$12.4 million) versus Caterpillar's profits of US$1 billion, which was 100 times its turnover. Hence, partnerships were necessary because it could not take on such a large player by itself. He commented: 'From a business perspective this collaboration is a must: otherwise we will vanish into thin air'.

Also, while having more collaborations boosted short-term results, he cautioned that success depended on alignment of partners' interests and a long-term orientation, particularly in R&D where rewards were reaped only after 10–20 years.

While growth was a key objective in India, overseas expansion was limited by the joint venture. Although Earthmovers' products were exported to countries like South Africa, Iraq, and Bangladesh, an executive noted:

> We cannot really decide many things without their consent and [Partner X] being a very international organization they have not really allowed us . . . to market these products elsewhere in the world besides the SAARC[26] countries or . . . certain countries in Africa . . . or in the case of the Middle East where there is a tremendous amount of construction activity. . . . So we use some of our channels to sell our Indian equipment but that is with the prior knowledge of our partner.

Consequently, the partnership constrained international expansion. Instead of direct representation overseas, attempts were directed to becoming a part of the Japanese partner's global network so the partner would source components from India. In turn, this would allow the Indian firm to enhance its capabilities and eventually address some markets with Indian products. See Tables 5.1 and 5.2 for a comparison of the evolution of the three firms.

ARCHITECTING COMPLEXITY IN OTHER FIRMS

The transformation undergone by Steelworks, Bearings, and Earthmovers in their attempts to expand and become serious global contenders was not unique. An examination of other manufacturing firms in the automotive and auto-components sectors, and observations from a leading industry association, suggest that other leading firms in this sector were responding to the challenge of global competition by adapting through restructuring and learning to employ organizational and technological best practices. New foreign competitors in the Indian market such as a minerals company providing services on a turnkey basis also experienced a similar process of learning. To match foreign competitors it became imperative to reduce costs

Table 5.1 Stage III evolution of three manufacturing firms

	Steelworks	Bearings Inc.	Earthmovers
Adopting modular organization	Mid-1990s reorganization in stages via: - Focus on productivity, customer satisfaction & quality - Focus on reducing hierarchy, introducing modular units linked via financial parameters and quality standards	- 1996 reorganization to create a more integrated system across borders - Matrix structure adopted but not uniformly - Training and overseas visits to ensure tighter integration with MNC headquarters; use of common systems for data gathering	- Began modularizing organization in mid-1990s - Adopted modularization practices from Japanese alliance partner, which included a focus on problem-solving and monitoring productivity, efficiency, and quality
Facilitating knowledge decomposition	- Adoption of IT - Establishment of standards & norms – disseminated via vision, mission, and strategic goals - New rules resulted from strategic assessment in every domain of activity, e.g. 'become EVA positive' - Outsourced peripheral activities such as town management	- Focus on increasing specialization and using India as one of the bases for global manufacturing and R&D Adoption of automation to standardize all tasks and equipment - New rule: 'quality at the lowest cost' for sourcing supplies; emphasis on logistics - Outsourced peripheral activities and even some grinding, turning, and heat treatment operations	- Adoption of IT throughout organization - Use of computer-aided design (CAD) and finite element method (FEM) for design - New product development simplified and speeded up - Financial orientation and emphasis on logistics - Outsourced peripheral activities
Real options heuristics	- Enhanced aspirations – Portfolio concept used to evaluate businesses - New products/services;	ROH strategy limited to: - Integrating Indian subsidiary into global organization	- Enhanced aspirations - Alliances to manufacture specialized equipment;

Table 5.1 (continued)

	Steelworks	Bearings Inc.	Earthmovers
	leveraged partner capabilities to produce a desulphurizing compound; managed port operations with a partner; pigment manufacturing; production of value-added steel - Input of R&D in strategic decisions to pursue innovation; emphasis on knowledge management to innovate; use of industry and academic consultants & experts - M&A to expand overseas	- Specialized in manufacturing some components; some R&D and new product development. - Expansion within India - Focus on leveraging opportunities and taking advantage of skills in a new location	evaluation of new business streams with a view to becoming a full line company rather than just an excavator manufacturer - ROH strategy constrained by alliance partner – integration as a supplier within alliance partner's global network; nevertheless, possibility of growth via M&A

Table 5.2 A comparison of the evolution of three manufacturing firms

	Steelworks	Bearings Inc.	Earthmovers
Mode of technology acquisition	Direct import	Joint venture (JV) with US firm	Technical collaboration with Japanese firm
Evolution of relations with technology supplier	Remained autonomous	US firm acquired Indian partner's share and made the Indian company a subsidiary	Japanese supplier became a JV partner by acquiring a 40% stake in the partnership
Location of expansion	Overseas expansion: South-east Asia, United Kingdom, Ukraine, Africa	Within India and possibly Asia; integrated as supplier within MNC's network	Within India and other selected SAARC countries with partner's knowledge; operates as node of Japanese MNC's network
Location of knowledge	Located in the firm and transferred overseas	Local manufacturing capabilities and some R&D capability	Technological knowledge located in Japanese partner; however, local adaptation capabilities and market knowledge exist in Indian JV
International expansion via:	Hierarchy; growth via acquisition	Hierarchy	Partnership
Aspirations	Leadership in global community. Now among the leading global steel companies	Expand global operations in India	Maintain leadership in India; integration in Japanese partner's supply chain
Community creation	IT-based; growing global community	IT-based; local linked with global community	IT-based; growing global community

Table 5.2 (continued)

	Steelworks	Bearings Inc.	Earthmovers
Functional area emphasis	More emphasis on customer orientation and finance	More emphasis on productivity and efficiency	More emphasis on customer orientation and market
Challenge	Rapid growth/HR-related Managing an operation that spans many countries	Rapid growth/HR-related	HR-related

and improve quality and productivity; attaining these required flexibility and modularity in organization, and adoption of new technologies to aid knowledge decomposition and codification. Finally, firms adapted their aspirations as they learned how to compete overseas and used a real options heuristics strategy to leverage capabilities, innovate, and expand overseas.

Adopting Modular Organization

In the case of a leading INR60 billion (US$1.5 billion) automotive manufacturing group focused on LCVs, utility vehicles, and SUVs, manufacturing technology evolved from adopting engine technology via a French collaboration to indigenous SUV design and manufacture. In the process, the organization was transformed to accommodate greater flexibility in manufacturing processes by deploying technology to obtain and codify knowledge.

Likewise, a leading manufacturer of trucks, buses, and automotive components underwent a similar restructuring to facilitate flexibility in the 1990s. Initially, equity investments by a British automotive firm brought new technology. The British investment was replaced by a joint venture between an Italian conglomerate and an overseas Indian transnational group in the mid-1980s. The advent of foreign competition in the 1990s led to organizational restructuring to flatten the hierarchy and enhance competitiveness. Most of the company's employees are shareholders; restructuring and switching from a top-down to a bottom-up approach made it possible to reduce costs by allowing 'everyone . . . a voice in terms of what should happen to the company in terms of quality, in terms of lean manufacturing, lean management'.

Facilitating Knowledge Decomposition

As in the case of Steelworks and Earthmovers, knowledge decomposition enabled every process to be monitored, evaluated, and controlled, resulting in deskilling labor.[27] An interviewee in the automotive manufacturing group noted that:

> We have to deploy technology and remove the skill away from the workman, get it into the equipment, and get it into the process. . . . Instead of doing higher-level testing of the engine . . . get into processes where we are able to map the various characteristics of the engine. The exhaust wave front is a good indicator of whether the engine has been built properly. So you deploy exhaust wave front checking equipment [and obtain information on] oil pressure, the pressure in the pistons, the injection . . . all these things you simulate so that you will not really need to test the engine by hard firing. . . . All these are ordinary equipment,

> which is very easily available because a lot of electronics has got developed where all these kinds of process parameters you can pick up, standardize them with a given sample engine, which is accepted totally in the engine testing area where you do cold firing to establish the torque required to turn the engine and then use that as a comparative value for fast produced engines. So this . . . increases the repeatability and reproducibility of the engine and reduces variability.

In addition, the firm also introduced CNC (computer numerical control) flexible machining centers instead of special purpose machines to avoid large investments in capital equipment that would not accommodate engine design changes.

Besides using technology to facilitate lower-cost production of new products, this firm also designed and developed a new SUV for the local market, innovating by recombining existing technologies[28] and outsourcing component production.[29]

The organizational transformation also led to a greater focus on knowledge management after 2000 in the truck manufacturer. This meant not only increasing emphasis on R&D but also adopting best practices in every area of the company. Such practices ranged from focusing on the stability and reliability of the product and fostering internal product design and development capabilities; the use of more sophisticated machine tools and material handling equipment; evaluating alternate materials to promote value engineering on the materials front; and adopting modern systems and procedures in finance to reduce financial costs. As in Earthmovers, the company had its own CAD/CAM[30] center although independent CAD/CAM centers were also available to outsource design. Thus, as in other firms, there was great emphasis on codifying all processes and activities to reduce costs and enhance quality. An interviewee noted that systems and procedures in Indian automotive companies and component manufacturers were not inferior to those in companies in Europe and the United States.

Quality certification programs played a major role in adopting information technology and standardizing processes. An executive at a national industry association noted that the association had helped to disseminate new management practices and quality standards (via ISO certification for manufacturing firms) thereby facilitating standardization and knowledge decomposition in member firms throughout India (see Figure 5.1 for the number of Indian companies adopting quality certification, and Appendix A, Table A2 for quality certification in the auto-component sector).

Real Options Heuristics

The advantages of lower cost, access to the local market, and the ownership advantage conferred by the new product gave the automotive manufacturing

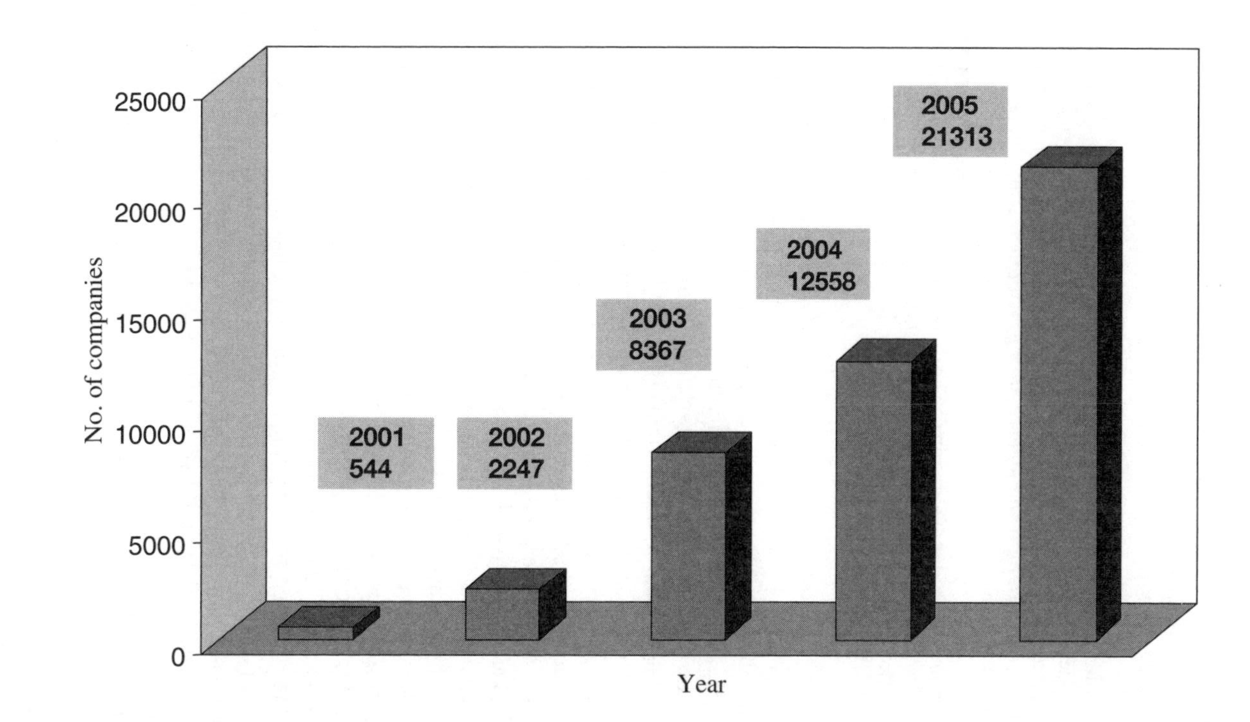

Source: ISO Survey 2004 and QCI (Quality Council of India).

Figure 5.1 Industry standards: number of companies certified by Indian certifying bodies from 2001–05

group a stronger position in negotiating alliances with foreign partners seeking access in India and fuelled new aspirations for global expansion.[31] While realizing shortcomings in R&D, top management understood the source of the firm's advantage in the automotive industry's global value chain and how its activities would be distributed across the globe:

> We are definitely not as advanced in product development. . . . Indian companies are strong in developing them at much lower cost . . . we can do it in one-third the cost. The multinationals are going to concentrate on designing of products because that is their strength and manufacturing is not their strength; neither will the cost structure in those countries allow them to manufacture vehicles. So, manufacturing for the next 10–15 years will shift to India. But the designs will be done abroad . . . computer-aided evaluation can be done in India but performance optimization, right handling of vehicles, driving comfort, and cornering stability . . . these need a lot of wisdom . . . this is where we have gaps . . . we will still take another five to seven years.

A real options heuristics strategy was implicitly followed as evidenced by the establishment of a joint manufacturing facility with capacity of 800 000 in India in an alliance with a French company and plans to set up an assembly plant in Brazil in partnership with a local company.

Similarly, the truck and bus manufacturing firm also revised aspirations upwards after acquiring a Czech firm. It articulated the desire to 'grow by acquisition wherever possible' and build capabilities in specific areas through alliances. Top management also began to consider the impact of external conditions on market creation and realized that the transportation system and infrastructure would have to change to meet the demands of an evolving market. Executives hoped to strengthen the ability to compete globally by raising the firm's R&D capabilities, focusing on innovation, and complying with worldwide emission norms. The company had already adapted to changing norms within India and was the largest producer of CNG (compressed natural gas) buses. Thus, building capabilities and mastering new rules of global competition via interactions with global players to acquire ownership advantages enabled the firm to enter global markets. Moreover, the presence of many major global automotive manufacturers and auto-component suppliers in the region together with strong local automotive component manufacturers indicated that existing capabilities could be leveraged further and that the region could emerge as a major hub for the Indian automotive industry.

The outlook for the region as an emergent location for manufacturing excellence was corroborated by the automotive component manufacturers association. The association's report indicated that OEMs (original equipment manufacturers) and tier-1 suppliers were beginning to view India as a sourcing base for auto-components such as transmissions, truck parts, two wheelers, engine parts, castings, forgings, plastics, rubber and electrical parts (see Appendix A).

However, new capabilities need to be developed continuously for locations to remain competitive. An executive of a leading industry association serving the automotive and manufacturing sectors observed that cost and quality were taken as given and that the only basis for competition was innovation.[32] Competition had increased owing to the availability of low-cost Chinese imports that flooded the Indian market. Thus, Indian firms faced the challenge of learning how to increase capacity rapidly while remaining cost-effective. Industrialists also realized that it was necessary to do higher value-added work and focus on new product development. This industry association targeted innovation as the new mission for manufacturing companies; professors from the Harvard Business School and MIT were invited to lead discussions popularizing the conception that: 'You can learn how to do innovation. It is a skill. That is what we are learning from experts on innovation'. It was hoped that disseminating new ideas and methods about innovation in manufacturing firms around the country would enable Indian firms to improve capabilities and compete on the basis of product innovation and differentiation.

As noted by executives of Steelworks and other leading Indian firms, the industry association executive confirmed that industrial expansion was accomplished mainly through a 'mergers and acquisitions' strategy that included acquisition of foreign companies overseas or locally by Indian firms and vice versa. Such moves were likely to lead to greater consolidation. Thus, it appeared that firms were following a real options heuristics strategy for expansion via staged investments. Expansion was especially evident in the auto-components and pharmaceutical sectors, both of which had been designated a success by the government at growth rates of 40–50 percent. However, growth rates in services surpassed those in manufacturing; among services the highest growth was experienced in the telecom sector at 60–65 percent. The interviewee also confirmed the conclusion of other executives that talent was in short supply because of the high demand for skilled labor. The concern was that rising salaries could reduce the low-cost advantage of Indian firms. Consequently, the planning commission had begun to focus on improving the skills and knowledge of the workforce with a view to enhancing employability in the manufacturing sector.

CONCLUSION

In conclusion, despite differences in the growth trajectory of firms there are similarities in how they evolved to achieve greater complexity. Each firm adopted modularity as an approach to organization and management and transitioned from functional specialization to modular specialization to reduce uncertainty and cater to increased complexity of products and environments. Each firm also used information and communication technology to decompose knowledge and partition the organization into sub-assemblies and components to yield greater flexibility and enhance the possibility of rational decision-making by reducing dependence on any particular group or individual, suggesting a rediscovery of Taylor's notions of scientific management. All organizations attempted to create a 'reliable' or robust system by incorporating redundancy via training, culture, and operating practices. While the transition to a completely modular system was not perfect, many sub-systems had been created in each of these organizations and the rules for integrating them had been learned and incorporated into organizational routines as evidenced by the use of financial analysis for managing all businesses in Steelworks. The discourse of strategic management permeated all organizations, suggesting that matching global competitors by increasing scale also dictated assimilation of new organizational techniques and culture.

Differences between firms stem from initial conditions. In the case of Steelworks, the transition was triggered by the adoption of new technology but autonomously managed and controlled so that adaptation and innovation were indigenous. The sense of ownership and achievement derived from managing major technical projects successfully led top management to pursue an independent path to commercialization rather than via collaboration with other organizations. The transition to a more global organization required a greater emphasis on strategy and R&D, particularly as it was critical to assemble new knowledge rapidly and find ways to transform and create a new value chain. Thus, the firm was quick to use external resources to learn new rules for operating effectively in a competitive global industry. A sense of entrepreneurship pervaded the organization and aspirations of leadership emerged through articulating a vision, evolving strategy and interacting with other leading global organizations in the industry as evidenced by the ability to use a real options heuristics strategy for domestic and international expansion.

In Bearings, on the other hand, operations had stabilized and the focus was on optimizing the global value chain. Hence, the MNC perspective dominated Indian operations both in the manufacturing plant and in the newly established R&D center in Bangalore. The insight that the real value of the Indian location was the availability of local talent for innovation led to efforts to leverage their position to expand into new Asian markets. Thus, although the strategy was similar to that of Steelworks, expansion efforts were directed towards growth within India and towards using India as a source for manufacturing, back office services, and R&D.

The perspective was one of integrating the Indian entity into the transnational organization.

Finally, Earthmovers' evolution as an independent organization was supported by the Japanese joint venture (JV) partner. Despite autonomous capabilities in design and new product development, the dominance of the Japanese JV partner in R&D and quality led the firm to rely more heavily on the partnership for technology. Moreover, as in the case of Bearings, which became a subsidiary of a multinational, despite strong growth and market leadership in India, the JV with a Japanese partner inhibited market expansion outside India with the exception of exports to some emerging economies. There was also concern about longer-term technology development since R&D spending was low in comparison with global firms. The firm appeared to be in the early stages of autonomous development in India while seeking to be integrated into its partner's value chain. Although continued dependence on the Japanese partner, both for technology development and market expansion, could be perceived as a source of vulnerability, new rules such as the use of mergers, acquisitions, and alliances for expansion were learned by interacting with foreign partners and other collaborators.

In summary, the leading Indian firms in the automotive, automotive component, and heavy equipment manufacturing sector appear to be adopting modular organization and using information technology to engage in knowledge decomposition and streamlining peripheral activities by outsourcing. In turn, these measures enable them to use a real options heuristics strategy for growth and expansion in India and abroad. While the widespread dissemination of best practices indicates that imitation of modular organization, knowledge decomposition, and a real options heuristics strategy by follower firms may be common, the extent of adoption and success are likely to be determined by aspirations, contextual contingencies, and partnership agreements, as well as the presence of resources and capabilities.

NOTES

1. I use Simon's (1956) definition of complexity.
2. The dispersion of the firm's value chain across locations is an indication of increased complexity as the firm copes with the need to gain flexibility to adapt to multiple sources of uncertainty (Kogut and Kulatilaka, 1994) as also noted by Ghoshal and Bartlett (1990) in their description of multinationals as networks. The evolution of such networks indicates the ability to deal with complexity.
3. The focus on benchmarking, cost and quality consciousness, customer orientation, and market orientation are common in the business literature and the popular business press.

4. See Chapter 2, endnote 36.
5. The total knowledge of the firm with respect to any particular technology is composed of knowledge that can be decomposed and recombined to create new knowledge and knowledge that cannot be decomposed (or articulated, according to Winter, 1987). Here I am concerned with knowledge that is capable of being decomposed.
6. Just as problems can be decomposed into sub-problems, knowledge can be decomposed and recombined to create new knowledge.
7. Holland et al. (1986, pp. 1–66) note that such parameters in a system can be viewed as rules or building blocks of the condition-action type: IF such and such is present, THEN do so-and-so. A cluster of rules or category that is evoked for problem-solving can be considered a default hierarchy. When the organization is partitioned into units governed by rule clusters it is possible to use building blocks that create combinations and linkages to generate new rules.
8. Again, these are rules in the sense of Holland et al (1986).
9. See Surie et al. (2003) for an elaboration of real options heuristics and the knowledge decomposition process.
10. Chandler (1962) notes the importance of consultants in diffusing the multidivisional structure overseas.
11. Holland et al. (1986) note the difference between generalized rules and specialized rules.
12. Nonaka (1994) notes that knowledge recombination is critical for innovation.
13. Innovation is an important factor in achieving minimum efficient scale, critical in capital-intensive industries (Audretsch, 1991).
14. Since its founding, the entire township had been developed and managed by the company. Water management was outsourced through a joint venture set up with an Italian firm, Vivendi.
15. Greenfield projects were planned in India, Iran, and Bangladesh. Growth through greenfield plants was estimated to be slow.
16. By industrializing knowledge production, accelerating the innovation process, forming linkages with other firms in the global network and assuming the role of a supplier, Steelworks was able to enhance its reputation and increase the scale of its operations domestically and overseas.
17. The Korean steel firm, POSCO, was also interested in locating manufacturing in India because of the lower-cost advantages of this location in comparison with Korea.
18. While this is evocative of Ghoshal and Bartlett's (1990) multinational (MNC) network, I focus on outlining the dynamics of network evolution here.
19. Improved trust and knowledge flows in interactions across borders in subsidiaries is predicted by studies on technology transfer (Davies, 1977) and others (Kogut and Zander, 1993).
20. Holland et al. (1986) note the importance of triggering conditions that lead to modification of existing rules or the generation of new ones.
21. Ross and associates note the errors made by the intuitive psychologist. Holland et al. (1986) reiterate the idea that statistical reasoning is not used in many circumstances.
22. Here the country subsidiaries and plants are viewed as the real options within the MNC framework (Kogut and Kulatilaka, 1994).
23. Compound annual growth rate.
24. Deliveries did not take more than two or three days.
25. Salaries were growing at 10–11 percent per year for managers and 7–8 percent for operatives.
26. South Asian Association for Regional Cooperation.
27. The attempt to deskill labor through automation is reminiscent of earlier studies on the effects of automation (Braverman, 1974).
28. Audretsch (1991) notes that new firms' survival is dependent on achieving minimum efficient scale and that scale is propelled by innovation.
29. See Appendix A for details on the development of the auto-component industry supply chain.

30. Computer-aided design/manufacturing.
31. This is reminiscent of Hymer's ([1960] 1976) finding that ownership advantage spurs globalization.
32. For the interview, the interviewee had to be taken out of a seminar on innovation that she had organized for association members; Clayton Christensen, Professor of Business Administration at Harvard Business School, was a keynote speaker in the seminar.

6. Industrializing knowledge production via born global firms: biotechnology and software

The two previous chapters discussed three stages of evolution of firms in the manufacturing sector and their progression from learning via transferring technology from external sources to institutionalizing the knowledge gained, and, finally accelerating knowledge production in the new environment. In contrast, in this chapter I focus on the rise of Indian firms in knowledge-intensive industries such as software and biotechnology. The manufacturing firms differ from software and biotechnology firms in three ways. First, while complex technical knowledge was required for the former (firms producing steel, bearings, earthmoving equipment, automotive components, engines, farm equipment and commercial vehicles), the knowledge underlying these technologies was well understood and diffused. In contrast, the biotechnology and software industries are less mature than steel and automotives, the underlying knowledge more complex, and embedded not just in firms but within an institutional system. Second, manufacturing firms were only able to convert to using information technology for all activities in Stage III. In contrast, software and biotechnology firms, being younger and being IT-based could modularize activities from the start. Third, manufacturing firms were isolated from global markets until the 1990s; consequently, the attainment of Stage I took longer. It was only in the 1990s that managers in these firms realized that it was not necessary to reinvent the wheel. However, the first two stages were accelerated in the software and biotechnology firms, because of their late entry and because they were interacting from the outset with global clients and were not reliant on the pace of development of domestic companies. Consequently, they are designated as 'born global' firms.

This chapter focuses on Stage III in the biotechnology and software firms. Leading Indian software firms are now recognized as global participants. In addition, many biotechnology firms were established. Some have already begun to gain recognition especially in bioinformatics and contract research services. Despite late entry into this industry, legislative change in the 1990s, founding conditions, and lack of administrative heritage enabled these firms to accelerate global participation. Radical technological change and innovations such as global project organization enabled firms to participate in global knowledge production by applying real options heuristics to leverage resources and technological capabilities.[1] An analysis suggests that expansion of both Indian firms and multinationals in biotechnology and software was stimulated by the need to accelerate innovation. All these firms used new technologies to facilitate to foundation of cross-border communities and flows of tacit knowledge. Consequently, geographic collocation was substituted by cognitive collocation. Thus, despite differences between the manufacturing and knowledge-intensive biotechnology and software sectors, firms in both sectors use similar processes for global expansion and growth.

This chapter first summarizes how knowledge production is industrialized by accelerating knowledge codification via the use of information systems and resulting changes in strategy and structure that accompany industrialized knowledge production. Three factors enable Indian firms to participate in global innovation:

1. extensive use of information and communication technologies (ICTs) and artifacts to automate knowledge production by creating knowledge components;
2. adoption of *heteromorphic* organizational form, which encompasses a variety of organizational forms over the technological life cycle to leverage knowledge;
3. use of *real options heuristics* (ROH) in the arena of knowledge production to reduce uncertainty via learning and gaining new information, and balance exploitation and exploration.

By codifying and recombining the knowledge underlying both technology and organization, complex organizations could be constructed using simple building blocks and cross-border expansion accomplished more rapidly. Moreover, such building blocks or intermediate components and products could be leveraged to gain access to complementary resources using real options heuristics as a strategy.

Next, the three-factor framework above is used to present a detailed examination of the global expansion of four Indian biotechnology and software firms (see Appendices B and C for a note on the historical background and evolution of the software and biotechnology industries). Figure 6.1 shows the positioning of these four firms based on their level of participation in the global community of knowledge-intensive firms and extent of geographical expansion. This is followed by a section on other firms, corroborating findings from the four cases in a total of ten other software and biotechnology firms[2] and seven biotechnology- and software-related organizations.[3]

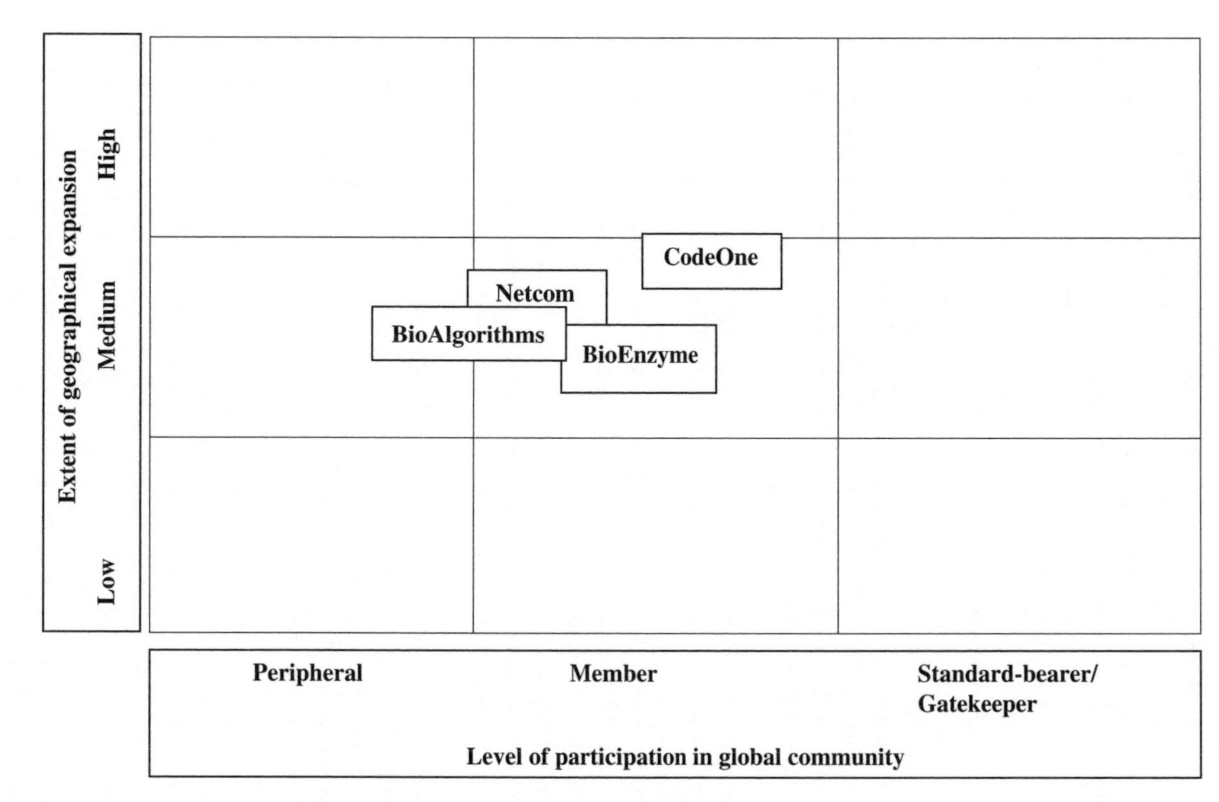

Figure 6.1 *Positioning of four Indian software and biotechnology firms in the global community of knowledge-intensive firms and extent of geographical expansion*

INDUSTRIALIZING KNOWLEDGE PRODUCTION

The rise of global biotechnology and software firms exemplifies the emergence of knowledge-intensive Indian industries. Such firms are characterized by their ability to leverage existing capabilities and accelerate knowledge production while continually seeking to explore and enhance future capabilities. Wider availability and improvements in technologies that facilitate codification have facilitated participation in global innovation. As codification converts knowledge to information that can be instantaneously reproduced, it permits outsourcing of activities (Cowan, David, and Foray, 2000). New codification technologies, tools, and artifacts that enhance cognitive learning and aid communication and coordination across organizational boundaries enable faster recombination of tacit and explicit knowledge and accelerate knowledge production. Codification thus stimulates the production of 'knowledge components' and allows firms to transact and exchange knowledge. As a result, firms can acquire more knowledge than before for a given (but not necessarily lower) cost (Cohendet and Steinmuller, 2000).

Moreover, the use of automation to create knowledge components through knowledge recombination allows firms to automate and industrialize knowledge production and accelerate innovation by increasing the number of experiments to fill the innovation pipeline[4] (Chandler, 1990). Consequently, organizational reconfiguration is necessary to accommodate a new division of labor resulting from outsourcing activities in which the firm has no specialization advantage (Cohendet and Steinmuller, 2000). Thus, craft-based 'guild' or master-apprenticeship systems that rely on intensive social interaction and task involvement in collocated epistemic communities to facilitate the transfer of knowledge are abandoned in favor of 'industrialized' systems that rely on technology to facilitate knowledge-sharing in geographically dispersed contexts.

Innovations such as the internet and new communication technologies also reduce coordination costs. Likewise, software techniques such as computer-assisted simulation and modeling promote rapid knowledge codification and dissemination (Cohendet and Steinmuller, 2000; Nightingale, 2000). Although 'uncodifiable' knowledge may exist (Cohendet and Steinmuller, 2000; Cowan et al., 2000), I focus on knowledge that, in principle, can be codified (Nelson and Winter, 1982 note the importance of incentives in rendering knowledge codifiable).

Greater organizational flexibility is required to allow firms to pursue multiple opportunities arising from the rapid recombination of knowledge and acceleration of knowledge production and innovation. Also, industrializing knowledge production requires speedy absorption even when specialized knowledge exists outside the boundaries of the firm. Consequently, the flexibility of the organizational system is enhanced by making all processes information-based and computerized, and designing the overall system to allow multiple goals to be pursued simultaneously using the most appropriate organizational form. Thus, firms can use intermediate forms like alliances to learn and acquire knowledge from external sources, or contractual arrangements to outsource non-core activities, while hierarchies can be used for commercializing innovations. This heteromorphic organizational form is characterized by the concurrent use of multiple organizational forms.

Finally, uncertainty in new technologies and the high risk and cost of innovation incentivize firms in developed countries to seek ways to reduce uncertainty and lower costs by participating in cross-border knowledge production. In addition, opportunity-seeking motivates multinational firms to seek partnerships with firms in developing countries that possess the requisite skills and capabilities to participate in knowledge production. Similarly, firms in developing countries seek cross-border partnerships to aid overseas expansion and compete in domestic markets by enhancing their technological capabilities. A real options heuristics (ROH) strategy helps to reduce uncertainty while allowing firms to pursue growth in knowledge-intensive industries opportunistically.

CODEONE INC.: CREATING THE KNOWLEDGE ASSEMBLY LINE

Founded in the mid-1980s, CodeOne was a software company that at the outset provided maintenance and other services to overseas clients. In the late 1990s, its workforce consisted of approximately 4000 employees; in 2006, the company had recruited 25 000 new employees and planned to add another 30 000 in 2007. The company had graduated from application development and maintenance to customized services in six horizontal areas such as certified solutions, infrastructure management and product engineering, and high-end work such as consulting.

Automating Knowledge Production Via ICTs: Generating Intellectual Property

The gleaming glass and granite architecture of CodeOne Inc. in Bangalore was reminiscent of Silicon Valley. Architectural replication was matched by the use of state-of-the-art information and communication technologies (ICTs). As a pioneer of the global delivery model, CodeOne had successfully used information technology to create a geographically distributed project-based workforce. ICTs were vital to work and taken for granted by managers and effectively linked organizational members with overseas colleagues, enabling round-the-clock information-sharing.

The industrialization of knowledge production is made possible when the production process is mechanized to enable handling large volumes of knowledge inputs or components. This, in turn, necessitates the acceleration of knowledge codification to enable rapid knowledge recombination for the production of knowledge components and knowledge modules or sub-assemblies to be used in the production process. Moreover, the production process need not be confined to a single firm but may be distributed across firms as some firms specialize in producing knowledge sub-assemblies, components, tools, or services for other firms.

Support was provided on a 24-hour and seven-day basis for outsourcing services for overseas clients, mainly in information technology projects involving system integration and custom application development, while basic services included maintenance and support. Project teams were coordinated across the globe and almost every project consisted of both 'on-site' and 'offshore' components. However, for economic reasons, 70 percent of the work was conducted in India. For example, typical offshore consulting projects involved working closely with the CIO (Chief Information Officer), the client team, and its partners, and client steering committees, which consisted of key client stakeholders and a representative from CodeOne. Work was partitioned into discrete chunks and distributed across the globe to India or China depending on where the competency was located. An executive noted that he worked with 150–200 people in the United States and 600 people offshore in one such project.

Having achieved close to US$1 billion in revenues in 2003, CodeOne executives looked forward to expanding rapidly overseas to match their competitors via innovation and noted that in the IT services industry: 'Today's competitive advantage is tomorrow's commodity. The only competitive advantage is innovation'.

Slow acceptance of the concept of cross-border work, stemming from the CIO's inability to understand how 'offshoring' worked and caution because of the perceived risk of doing business offshore, resulted in the dominance of maintenance-related work during the early years of the company. However, over time many overseas clients had come to realize that offshore work was necessary for their own survival. Consequently, executives focused on how to provide value: 'Can we standardize the process to make it repeatable . . . standardize it and stay ahead . . . make the money and move on to make the next thing?'

Even though the company experienced rapid growth, top managers were aware that it was important 'not just to add more and more people at the back end' but to differentiate and add value to clients. Differentiation was particularly important as other competitors had started emulating the model; hence, retaining a competitive edge involved ensuring both rapid growth and differentiation. Consequently, the emphasis on differentiation led to a focus on R&D and entry into areas like consulting.

Knowledge production could be accelerated by leveraging technology to source knowledge resources. The experience gained in conducting offshore IT-related projects had enabled CodeOne to develop an advanced knowledge management system that allowed project managers and executives to access and leverage existing knowledge resources and assets stored according to a defined knowledge hierarchy. For example, executives and managers could access past projects and identify key people relevant to their current project and thus gain access to both tacit and explicit knowledge:

> At the completion of a project or during the course of a project there are different knowledge artifacts which get created by each of the project teams. These knowledge artifacts are then stored based on a defined knowledge hierarchy and with knowledge nodes or sub-nodes; each of them are mapped and housed in different servers which are all linked to the knowledge portal.

Consequently, available resources could be speedily marshaled to construct solutions. For example, the solutions consulting group focused on business problems and solutions to address those business problems that were 'repeatable' across an industry vertical such as retail. Likewise, as part of the initiative to innovate, the quality area focused on software automation and componentization to 'create software in an assembly line fashion'.

Information technologies thus enabled the acceleration of knowledge production and the generation and management of knowledge components and artifacts. These could be recombined and used across the company to solve specific problems in different domains, thereby reducing the time taken to arrive at a solution.

One solution strategy was to build prototypes or 'accelerators' either on the delivery or sales side instead of building a complete product. For example, a sales accelerator could reduce the sales cycle time from six months to two months because it showcased intellectual property built up over many projects and embodied in knowledge artifacts. Likewise, a delivery accelerator such as a process template or a component of code could help to speed innovation for clients. 'Master data management', another such solution was expected to yield revenues of US$15–20 million in an industry business unit with revenues of US$290–300 million.

Similarly, a multichannel commerce solution was built around knowledge of website design for retailers. Although the code developed for each customer's website was unique to the customer and belonged to the customer, general experience gained from designing multiple websites by individual architects and project and program managers could be received elsewhere. This included experience in estimating the effort required for such programs and converting it into a knowledge artifact (that did not violate the intellectual property agreement with the customer).

The creation of accelerators and knowledge components enabled the firm to adopt an industrialized mode of knowledge production and sustain a larger flow of innovation. Accelerators also allowed the firm to differentiate itself vis-à-vis competitors. Increased spending on R&D (2 or 3 percent of revenues) also allowed the firm to focus on providing software as a service or utility, an area that was expected to increase revenues by 10–15 percent in the next two or three years. Finally, the rising acceptance and active pursuit of cross-border work by clients indicates that geographical collocation may to some extent be substituted by cognitive collocation.

Evolving Complexity: Creating a Heteromorphic Organization

A key aspect of evolution in organisms is an increase in complexity (Pringle, 1957). Managers and executives were aware of the need to design the organization to accommodate increasing complexity while allowing for flexibility. Moreover, organizational complexity enabled the firm to match its strategy and environmental complexity.[5] Design encompassed both the internal structure of the firm as well as its relations with clients and competitors and, thus, extended the boundaries of the firm via formal and informal agreements with other organizations.

The global delivery model developed early in the firm's history as a result of working with overseas clients, used project teams that operated non-hierarchically to allow for information transfer between the client site and offshore project team and enable the distributed teams to function as a knowledge-sharing community. CodeOne could thus function as an extension of the client organization.

A hierarchical design was nevertheless adopted to achieve scale while facilitating extensive knowledge-sharing and rapid solution-building. For example, offshore divisional managers were responsible for multiple accounts and were supported by delivery managers, group project managers, senior project managers, project managers, technical architects, software engineers, and programmers. Overseas subsidiaries were organized by industry and geography. The US subsidiary was organized by industry; in contrast Europe, Australia, and Asia-Pacific were geographically organized, as the volumes to justify organizing by industry did not exist.

Despite the hierarchical structure, flexibility was maintained by using IT systems to deploy knowledge embodied in specific individuals wherever required (Hall and Johnson, 1970). Although each industry vertically housed its own technical experts, key resources could be accessed from other parts of the organization through the delivery system and allocated to a specific project. If the resource was already allocated to an existing project (utilization rates of 82–83 percent were common), it was, nevertheless, possible to acquire it for a new project depending on the project's priority. Boundary-crossing across vertical domains was not required more than 10–15 percent of the time; however, it was more prevalent in the consulting domain. For example, in the retail industry vertical there were 250 global strategy consultants and the resource pool was very 'fungible'. Technical architects formed another fungible resource pool as they were experts in a specific technology and could be deployed across different industries. Similarly, 'horizontal enterprise capability units' were common resource pools with expertise across a range of functions that were standard across all industries and could be tapped by each industry vertical based on requirements. Thus, individual units or modules within the hierarchy were linked together by experts or expert units that functioned as interfaces, helping to preserve flexibility. Typically, within a business unit such as retail, 25–30 percent of the revenues were accounted for by the enterprise capability unit, and about 70 percent through the industry delivery unit.

This flexibility was also built in by ensuring that training (through the company's extensive programs on enterprise solutions, domain enablement certification, and leadership training) was an integral part of an employee's career path. Thus, individuals were required to demonstrate proficiency in skills required for the next level before promotion, much as in the German vocational training system.[6] Consequently, by ensuring certification and creating a hierarchy of skills ranging from technical to cross-functional and leadership skills, unit heads were able to configure or disband project teams rapidly to suit clients' requirements.

Structural flexibility was also enhanced by creating and using a pool of certified subcontractors. Subcontractors were used relatively infrequently (and accounted for less than 10 percent of the business) to ease supply constraints and add capacity; subcontracting was mainly used for technology resources.

Additional evidence of organizational complexity was the prevalence of multiple alliances with other firms, including competitors. An alliance organization had been formed in 2003 to manage partnerships of all types: alliances were divided into three tiers. The first tier included global strategic alliances. These were alliances with key players in the industry from whom they could gain early information about industry trends and new products and also obtain training for key organizational members. However, global alliances differed from those of traditional system integrators in avoiding revenues through influence or license fees; instead, CodeOne focused on obtaining 'soft credits' such as 'training credits' for product and system training for employees. The second tier or non-strategic alliances were regionally based and the third tier included alliances with smaller firms or team arrangements for obtaining access to specialized tools such as tools for task management in the retail sector. While less than 5 percent of revenues came from alliances (on total revenues of US$1 billion) when the alliance organization was formed, current revenues from alliances were close to 15 percent (on revenues of US$3 billion). The importance of flexibility in maintaining competitiveness was clear: 'You need to build loyalty through innovation . . . You should be able to very quickly restructure, repackage, re-platform and re-gear your operations to produce new products or new services. The basis for competition would be new products and services'. Thus, organizational design encompassed a range of informal and formal mechanisms to extend the boundaries of the firm. Alliances were used to gain access to knowledge, new customers, to learn about standards, recruit new employees and influence curricula at universities and technical institutions.

Finally, by obtaining certification (SEI/CMM[7] level 5) CodeOne, like many other software firms, signaled its intent to comply with rules established by global players in the industry. Similarly, knowledge of the 'codebook'[8] was also obtained by recruiting staff with the relevant expertise in markets like the United States. These efforts helped to build the necessary interfaces between different units across organizational and national boundaries and ease the task of cross-border innovation.

Expansion Via Real Options Heuristics

Indian firms like CodeOne expanded by following a real options heuristics strategy. Such a strategy encompasses using knowledge components or services developed at various stages in the innovation process to gain access to new or complementary resources and capabilities. Knowledge created by the firm is treated as an asset or a real option that can be valued and exchanged for cash or other assets. By building on its initial offering of basic cross-border software services, the company was able to learn from clients and progressively enhance its offerings to the point where it began participating in high-end work and cross-border innovation.

CodeOne had dealt with international clients from the start and established its first US office in 1987. However, during its early history, the company largely used a labor arbitrage strategy using low-cost Indian labor to provide maintenance and other services. With the growth of revenues, an IPO (initial public offering) in the Indian market, ISO certification, and the liberalization of the Indian economy, CodeOne was poised for further growth and established development centers across India to support the globalization initiative as well as offices in Europe and Canada (1995–97).

Further quality initiatives were pursued and CMM level 5 certification was achieved in 1999. These, together with mastery of the global delivery model and increased visibility of the firm in Indian and foreign markets, enabled the company to enter new markets such as consulting and e-business and focus on packaged applications. By 1999, the company had achieved US$100 million in annual revenues, was listed on US stock market NASDAQ, and had offices in other European countries and Australia and two development centers in the United States. CodeOne then embarked on reorganization to accommodate new groups including domain competency, software engineering, and technology and communications. These efforts resulted in annual revenues of US$1 billion by 2004, the formation of a consulting subsidiary, and over US$2 billion in revenues, with more than 50 000 employees by 2006.

Evidence of the company's progress suggests that at each stage the firm made moves to accommodate market changes and leveraged existing resources to acquire either new capabilities or complementary resources. Moreover, as the company's position and profile improved, aspirations were continually revised upwards and new goals established. Thus, an executive in the United States noted: 'Three years back we were just an ADM [Application Development and Maintenance] body shop where people used to come and ask us for software engineers; today 40 percent of our revenues come from services which we created in the last three years'.

Although alliances were used to facilitate cross-border expansion, growth largely occurred via hierarchical expansion, mainly because the firm was able to raise sufficient resources through revenues and from capital markets.

NETCOM INC.: PROVIDING OUTSOURCED R&D FOR PRODUCT INNOVATION

Netcom, Inc., founded in 1989 by an electrical engineer with a Master's in computer science from the United States and 11 years' experience in Silicon Valley, focused on providing complete product lines rather than non-core types of outsourcing services and products to clients in the telecom industry. Consequently, the company interfaced directly with the clients' customers:

> For example, when we work with a large equipment manufacturer, we actually do a lot of development on their products which are already in the market. But

> [we work on] new releases which have to be put in, changes that happen because of changes in the underlying semiconductors or underlying operating system, or working with them on fixing bugs . . . interacting with their customers directly.

As in the case of CodeOne, key executives recognized the need to 'differentiate ourselves and add more value to our customers' and had begun to focus on leveraging intellectual property by developing innovative products to be sold to customers globally. Innovation was especially important as the company had relied mainly on internal financing; while there were other Indian competitors, they lacked the depth of experience in doing this kind of specialized work.

Automating Knowledge Production Via ICTs: Generating Intellectual Property

As in other software firms, ICTs were taken for granted and knowledge production was automated. Knowledge of the 'codebook' was acquired through interactions with overseas clients and by participating in standardization processes. Thus, by adopting the same technologies and standards used by overseas clients it was possible for members of the firm to participate as peers in cross-border innovation. Membership in overseas standards bodies gave the company early access to new ideas and policy changes while providing the opportunity to participate in the evolution of technologies:

> We participate in standardization processes. We're members of ITU [International Telecommunications Union], 3GPP [3rd Generation Partnership Project], GCF [Global Certification Form] technology bodies and of the ATM [Asynchronous Transfer Mode Forum], DSL [Digital Subscriber Line Forum] and SDR [Software Defined Radio Forum] forums. As a matter of fact, one of our ideas is an essential basis in DSL standards. The beauty of it [of participating in standardization processes] is that it's an engineer-to-engineer debate. A degree of objectivity prevails.

Top management was aware of the need to create intellectual property that would confer an ownership advantage. Consequently, the company had filed for 32–35 patents in the United States in 2003, and had successfully obtained nine patents by 2007:

> We work in the telecom segment and in that business unless you own patents you will not be able to compete effectively in world markets because most of this is driven by standards and these standards are essentially generated, driven from patents that the companies have come up with.

The company focused on high-end design work that resulted in intermediate products for other wireless and telecom companies. This included system and algorithm design that was then implemented and tested in live production environments. By working with semiconductor companies at the outset, Netcom was able to speed the process so that the only thing customers had to do was to select the silicon chip (which contained the complete system) for their final product. Although Netcom's products were components for their clients' products, and considered an 'ingredient brand', the company's goal was to make final products such as mobile terminals and smart phones. An example of the firm's capabilities in performing high-end work is data transfer at one megabit per second via cable for a US oil rig company while drilling for oil at a depth of 44 000–45 000 ft (13 000–14 000 m) under hostile conditions such as temperatures of about 200° C. Patents generated from this work would be jointly owned with the client with the legal agreement stating that Netcom would not use the technology for any other oil drilling company. Similarly, Netcom owned a patent jointly with a Chinese client.

Evolving Complexity: Creating a Heteromorphic Organization

Although Netcom had grown via hierarchical expansion, the company had alliances and extended relationships with firms in complementary industries. These alliances ranged from partnerships with clients with whom Netcom owned joint patents, silicon vendors, and other suppliers. Working with semiconductor vendors enabled the company to reduce the time taken to put its systems on to silicon chips. Likewise, Netcom also outsourced peripheral activities such as testing to reduce resource requirements. To ensure that proprietary intellectual property was not appropriated by other parties, the process was embedded in a 'black box'. In other instances, the technology developed in a partnership was patented and assigned to both parties.

Unlike other start-up firms, however, Netcom was a leading wireless telecom company in mobile communications. As a result of its involvement in high-end work and R&D for clients Netcom participated in standards bodies. Top management realized that:

> In the telecom space we're in, what's happening, interestingly enough, is that everybody believes you can never succeed in this market space any more unless and until the majority of the people follow the same standards. And bodies are coming up which are becoming intermediaries, arbitrators, etc, to ensure that there's nobody acting funny.

Participation in various communities such as standards bodies made organizational boundaries more permeable; this increased flexibility in design allowed the firm to acquire first-hand knowledge about industrial and technological changes and participate in shaping their evolution. Participation in standards bodies also provided an understanding of the overall architecture of the industry and thus helped in locating potential niches and positioning.[9] Top executives were aware that competitiveness involved developing networks of people interested in following a particular standard, thus reinforcing past research on the importance of networks.[10] Complexity in organizational design was consequently an outcome of the need to create interest in the new technology, generate vendor support, gain customer loyalty, acquire ownership of technology, and service markets in different countries. Complexity was evidenced by the variety of organizational structures concurrently in use. Despite the reliance on alliances for sourcing knowledge, global expansion was mainly achieved via hierarchies; sales offices and development centers and new subsidiaries for network solutions and product life cycle management were established in the United States, Europe, Canada, Mexico, China, and Japan.

Expansion Via Real Options Heuristics

Netcom followed a real options heuristics strategy by focusing on acquiring and leveraging ownership advantages conferred by automating knowledge production and generating intellectual property. By forging partnerships with leading players in the United States to obtain certification in key technologies and achieving SEI/CMM certification (level 5), the company was able to take advantage of new opportunities. The company also formed partnerships in specific sectors such as with emerging specialist semiconductor developers. In addition, depth of experience was gained by working with different types of customers in various countries.

By 2003–04, by focusing on different aspects of the value chain such as semiconductors, network equipment original equipment manufacturers (OEMs), and terminal devices, the company had established itself as a strong player each of these segments. The company was also increasingly recognized for its ability to deliver on complex projects to customers. The intellectual property side of the business had grown substantially since 1997 when the first DLS license was sold. In addition, a phone using the company's GSM/GPRS protocol stack had been launched in China and tested in several networks worldwide. Similarly, multimedia application suites were being used in handset models in Japan, China, the United Kingdom, Australia, and Hong Kong. To sustain the growth of intellectual property the company planned to increase R&D spending from 1 percent of revenues to 3 percent and continue to focus on patents. A separate subsidiary was also created to provide an offering to operators in 2003–04. In

2004, another subsidiary was created to provide installation, commissioning and project management services in the area of cellular networks.

As a result of the growth of revenues (by 45.5 percent in 2004–05 over the previous year with an on-site:offshore revenue ratio of 25:75) and the achievement of significant milestones, the company was able to attract new investments from leading clients in the telecom sector and an Indian venture capital company. In September 2005 the company launched a successful IPO. These infusions of cash allowed top management to pursue a mergers and acquisition strategy, which was implemented in 2006 with the 100 percent acquisition of a provider of wireless R&D and testing services in Finland.

Thus, it can be seen that Netcom pursued a real options heuristics strategy, either implicitly or explicitly and leveraged its resources at each stage to acquire new capabilities or assets. Like CodeOne, its revenue growth and resources permitted hierarchical expansion while partnerships were used mainly for knowledge sourcing. Unlike CodeOne, however, Netcom followed a product and intellectual-property-oriented strategy from the outset and was well aware of the need to patent and invest in R&D.

BIOENZYME INC.: MARKET-DRIVEN RESEARCH AND INNOVATION IN BIOTECHNOLOGY

Founded in 1978 as a joint venture with an Irish company to manufacture outsourced products such as enzymes using indigenously available raw materials, BioEnzyme Inc. was involved in manufacturing innovation and global markets from the outset. As the company did not have access to the manufacturing technology for these products in the United States and United Kingdom (despite availability of the technology), BioEnzyme developed the technology locally. However, in 1984, the Indian founder initiated an R&D program to manufacture new enzymes by obtaining funding through local grants, loans and revenues generated from product sales. The company achieved plant-scale technology and production enzyme successfully despite limited resources within five years of starting the R&D initiative largely as a result of focusing on market needs in pursuing research. A proprietary product and technology was developed and the company was able to obtain funding from an Indian venture capital firm. This coincided with the acquisition of the company's foreign partner by a European multinational, leading to recognition of BioEnzyme's contribution to high-value intellectual property generation in India. The emergence of a multinational partner with deep pockets gave BioEnzyme greater visibility and the opportunity to participate in contract research projects. Consequently, revenues increased both as a result of new contract research projects as well as royalties on products developed in India.

Automating Knowledge Production Via ICTs: Generating Intellectual Property

Like CodeOne and Netcom, BioEnzyme Inc. focused on creating value through differentiation from the outset and top management was very aware of the need for differentiation:

> We like to differentiate ourselves because we feel, long term, that's the best way to go because you'll always be much more competitive, and if we have to stay ahead of the market . . . . we have to do things differently because otherwise you don't create hurdles or entry barriers.

From the outset, ICTs were taken for granted as the company focused on generating products and technologies for its multinational partner. As in the case of the software firm, knowledge of the 'codebook' was developed over time via interactions with the joint venture partner, clients, and other alliance partners. In addition, knowledge recombination was facilitated by the presence of local expertise in diverse disciplines such as chemistry, microbiology, molecular biology, fermentation, chemical engineering, biochemistry, and pharmacology (the company had 460 scientists in 2003, of which 110 were PhDs). The company generated sufficient revenues from royalties on research done in India, demonstrating high levels of productivity and sustainability to their European joint venture partner. Top management and researchers were aware of the need to be 'patent literate' and develop an 'ethos of innovation'.

Having moved away from enzymes to pharmaceuticals, although the company focused on known molecules, proprietary knowledge was used to create novel products such as a bioreactor that was patented worldwide and thereby gain a strong non-infringing position. Likewise, the company also attempted to differentiate itself in the area of recombinant products to gain ownership advantage.[11] This included following a process in which DNA contamination was reduced to produce oral insulin. Similarly, through its clinical research subsidiary the company had developed knowledge about disease progression in diabetes and oncology, and had filed three patents on diabetic nephropathy.

Finally, the importance of gaining legitimacy by complying with the regulatory environment of the country in which the client was located was well understood. The company obtained certification from a leading US standards body for clinical trials conducted in its new laboratory.

Evolving Complexity: Creating a Heteromorphic Organization

Coordination and organization were critical for automating innovation and centered round creating an 'innovation ethos'. Thus:

> You have to have a very, very integrated team of people working at the same project. . . . In most companies they compartmentalize it, but it also appears very often that the process development group is a separate division, the lab development group is separate, the regulatory people are a separate group and each one is not integrated. . . . It's not a team effort to look at the target in a passionate way. . . . The commercial angle is also disconnected. . . . there has to be a pull from marketing, from R&D, from process development, from manufacturing, from regulatory, so everyone had to feel that this was a very, very important project.

To complement internal integration and accelerate innovation while retaining flexibility, the company used partnerships to gain access to complementary technologies.[12] Rather than attempting to do everything in-house, it was critical to consider how much value addition was achieved when in-licensing technologies. Examples include an alliance with a foreign start-up company that had developed an innovative technology for the food industry. Similarly, other platform technologies were licensed from many partners, including an alliance for early stage discovery and co-development of four therapeutic antibody products with a company based in the United States. As much of the research was conducted with the help of overseas partners the company was familiar with issues pertaining to the protection of intellectual property and professionalism in dealings with partners was taken for granted. The local adoption of standards and norms common in the United States and other industrialized countries was also essential in establishing trust in interactions with foreign partners.

Other mechanisms adopted to gain access to expertise included forming an advisory board, which included experts in key scientific disciplines and complementary areas such as patent litigation. Finally, despite limited resources, the company planned to maintain control over commercialization by expanding hierarchically and shift from being a producer of enzymes to becoming a biopharmaceutical company. This plan became possible with a successful IPO in 2004 that raised the company's market value to US$1.11 billion. New subsidiaries were also created for early stage discovery and development and clinical trials and customized research.[13]

Thus, it can be observed that multiple organizational forms were being used concurrently: knowledge sourcing was accomplished via formal and informal alliances at various stages ranging from discovery to development while commercialization was achieved via hierarchical expansion and the establishment of subsidiaries and sales offices in India and overseas. Each organizational form contributed to the achievement of goals. Consequently, a new firm lacking in resources could leverage existing resources and capabilities for expansion and entry into different markets and domains by means of alliances and achieve goals at interim stages until it developed the capability to expand autonomously.

Finally, the adoption of heteromorphic organizational form (H-form) from the outset facilitated the extension of firm boundaries, thereby enabling the firm to balance the contradictory goals of flexibility and control.

Expansion Via Real Options Heuristics

Although like other start-up organizations, BioEnzyme Inc. lacked resources to pursue basic scientific research, this did not preclude innovation. Since a key goal was to achieve differentiation through innovation, the company focused on radical innovations that could not be easily imitated. Examples include novel products such as a proprietary bioreactor (patented in 2001), and an innovative process to deliver insulin orally, which resulted in a new drug. Patent applications were filed (the company had over 100 patent applications in 2006) as products and technologies were developed, creating knowledge assets that allowed the company to enter strategic partnerships and extend its boundaries. In addition, the company exploited its knowledge base to generate revenues by offering contract research services. 'We want a very large market reach . . . we don't have that much pharmaceutical experience . . . we rely more on selling to large pharmaceutical companies or biopharmaceutical companies and getting them to market, and we just give them the active ingredient'.

Thus, at each stage, resources were leveraged to achieve new capabilities and aspirations were adjusted upward using a real options heuristics strategy. During its early history BioEnzyme Inc. focused on demonstrating its ability to conduct research to its partners and did so by attaining ISO certification in 1993. The same strategy was used to enter the clinical research area by gaining certification from the College of American Pathologists (CAP) for its laboratory in 2002. A market orientation was clear from the beginning; by incorporating a market perspective in product development, commercialization was easier as all areas of the company were able to contribute and understand the importance of the project.

The adoption of a real options heuristics strategy was also evidenced by the careful monitoring of projects that had to adhere to a strict timeline. For example, the target for oral insulin was to produce a prototype within one year, not five. Moreover, top management was also aware of the need to abandon the project if the requisite progress was not achieved within the target timeframe: 'One of the processes is to give up. I think we are much more flexible and much more adept if you take quick decisions. . . . Of course the danger is that you might take a certain impulsive decision'.

By 2004, the company had begun to manufacture its new bioinsulin and by 2006 had launched the first drug for cancer made by an Indian company. Adapting aspirations to a new context at each stage necessitated building innovation capabilities to match other global players rather than relying on a low-cost strategy:

> If I make [xxx] and look at it as an affordable drug, that's largely because I believe that it is a very strong marketing tool. But even if I had a more expensive way of making [xxx] I would still do it because I think that's a very important invention and that's a very important opportunity for us to innovate. To me, that serves me as a learning curve . . . If I can make a drug very cheaply and I can compete in the world market that's a strength but if that's my only strength then it's not a sustainable model long term.

BIOALGORITHMS INC.: INNOVATIONS FOR ACCELERATING R&D

Founded in 2001, BioAlgorithms Inc., a bioinformatics and contract research company was global from the start. By the beginning of the 21st century, a decade of experience with economic liberalization had created an environment in India that favored the development of knowledge-based organizations. The company began as a contract research organization with a laboratory in the United States and a development center in India to do contract sequencing for other firms and to sequence medicinally viable plants. However, a change in the patent laws preventing the patenting of sequences set the US laboratory on the trajectory of pursuing diagnostics while the Indian development center focused on developing products such as a sequence analysis product.

Automating Knowledge Production Via ICTs: Generating Intellectual Property

With offices in the United States and India and a 24-hour helpdesk to serve clients in the United States, ICTs were taken for granted and formed the communication and coordination infrastructure for BioAlgorithms, Inc. especially as the company was in the business of producing software to automate research processes in laboratories. Like the other companies, BioAlgorithms focused on generating intellectual property despite being a start-up firm with only 40 people in the Indian center in 2003. The company had created about five products for accelerating R&D in the biopharmaceutical industry. Although developing the sequence analysis product had been difficult, the experience had helped build a biological knowledge base in the company. In addition, the company's first two or three products were developed to keep track of information flow in the laboratory, resulting in the creation of a laboratory information management system (LIMS) that was compliant to Food and Drug Administration (FDA) guidelines in the United States. It was also customized for scientists in R&D laboratories unlike other available LIMS products that were adaptations of accounting or financial packages. Although global competitors existed, BioAlgorithms had a monopoly position in India.

Other products for the laboratory included micro-array packages for tracking and analysis, and packages for protein and structure prediction analyses. Besides automating laboratory experiments and other processes, knowledge was recombined for new software products. Modular software design allowed modules to be recombined and packaged in various combinations for different clients. The CEO noted that:

> We have all these products; for us it is very easy to do custom software. Recently we did one for another large US company. We created a niche product for them; it was essentially a package where we took two or three of our packages and customized and mapped their entire work-flow. So what they were doing with 10 or 20 different manual steps with different packages we just took and within two months we delivered them a solution that would suit their work-flow. And the good thing was we were actually competing with huge IT companies like [X] for this particular product. And we beat them on all grounds: cost, time, delivery, and understanding of the system. So, that's where our advantage is, because we have all the products in place. And we own the rights to all our products.

Besides marketing the products, the company also used them to conduct analyses for clients and had filed for patents for only some selected algorithms because of the expense involved in filing for all. In addition, the company also provided a knowledge management module that enabled clients to search and manage information. By early 2007, the company had graduated from bioinformatics to manufacturing micro-array chips and setting up ISO-certified laboratories to conduct experiments on a contract basis.

Evolving Complexity: Creating a Heteromorphic Organization

Software systems similar to those developed for pharmaceutical clients to help them track experiments in the laboratory were commonly used within BioAlgorithms to coordinate work and share knowledge generated by the group. The adoption of such systems was facilitated by the founder's familiarity with similar systems for sharing and managing knowledge in software companies in the United States. In addition, obtaining ISO certification also helped to coordinate and modularize work by standardizing processes. By establishing controls for the knowledge that was generated, work could be coordinated within and across organizational boundaries. Despite the initial difficulties of adopting a system to promote knowledge-sharing, strong incentives to do so were provided by setting the rules. According to the founder:

> Basically we said that if you don't use the system, you don't stay here. And slowly they got used to it. And now when anyone new comes in it is very easy for them to adopt it because everyone is using it. So, I think it is just a matter of setting the rules.

Standardization and modular organization allowed BioAlgorithms to function as an extension of the client organization and become an outsourced product developer. Thus, the company had engaged in product development for a bioinformatics company in Singapore that lacked knowledge in the areas of sequence analysis and micro-arrays. In order to build long-term trust and maintain the confidence of the client, norms common in market-oriented economies were followed. BioAlgorithms obtained no source code but received a baseline product from the client organization, which was developed into a more advanced product. However, the client was given the entire source code with the finished product.

Besides organizing to facilitate knowledge creation in a modular fashion, the need to gain access to complementary resources led to the formation of alliances in areas where knowledge was lacking internally. For example, in 2003, the company had also forged an alliance with a software company in the gaming and entertainment industry in Canada to provide algorithms for visualizing molecules for a new software package for protein analysis.[14] While BioAlgorithms would provide the bioinformatics expertise, the partner would focus on producing the images, and both partners would jointly own the intellectual property generated. Similarly, in 2004 an alliance was forged with a New York-based hospital for analyzing musculoskeletal diseases using BioAlgorithms' micro-array analysis tool. Other partners included leading global firms in the information technology sector.

Over the next three years, as the reputation of the firm grew stronger with large pharmaceutical companies as clients and as revenues increased, the company changed its focus from pure bioinformatics to conducting laboratory work. A consequence was the acquisition of two companies in Europe. The first acquisition was a division of a company manufacturing micro-array chips while the second was small company manufacturing oligoneucleotides. In addition, the firm had established a presence via a subsidiary sales office in the United States and through distributors in Taiwan, Israel, Japan, Italy, and Austria. Thus, adaptation from a small entrepreneurial bioinformatics firm to a life sciences company entailed evolving a heteromorphic structure to incorporate flexibility by using a variety of different forms.

Expansion Via Real Options Heuristics

BioAlgorithms had leveraged resources and acquired capabilities and experience rapidly in the short span of five to six years using a real options heuristics strategy. The CEO explained that the company was focused on bioinformatics, which required a learning period of about three years; the plan was to graduate to research-driven discovery after gaining the requisite skills and building a brand that was internationally recognizable, suggesting that the exploitation–exploration trade-off was well recognized (March 1991). As noted above, the company expanded rapidly from its inception in 2001 and leveraged its resources at each stage to gain access to new knowledge, clients, and markets. As in the case of BioEnzyme, during the early stages, the company focused on building legitimacy and credibility with customers via ISO certification, employee training, and adoption of common industry standards. These steps made it possible to attract one or two large pharmaceutical companies as clients initially,[15] which as the CEO noted, made a big difference when marketing to Indian companies.

Moreover, in the early stages, BioAlgorithms used resident expertise in information technology to provide services such as laboratory information systems and also outsourced product development. However, as resources increased, the company graduated from a purely bioinformatics company to manufacturing micro-array chips as top management recognized that survival as a bioinformatics company was unsustainable. The early strategy of internal growth and boot-strapping was replaced by growth via acquisition in 2006 when the company was able to obtain US$6.5 million in funding from the International Finance Corporation. Two acquisitions in Europe enabled the firm to acquire new clients and also acquire new technology in micro-array manufacturing and oligoneucleotides while a global presence was initiated via distributorships. These changes in strategy were accompanied by aggressive new targets for growth that involved growing the company from US$3 million in revenues to US$10 million in the next year. Key challenges involved in rapid expansion included the need to establish systems and obtaining and retaining the right talent to achieve aspirations within the desired timeframe. Thus, BioAlgorithms followed a real options heuristics strategy either implicitly or explicitly as evidenced by actions taken to leverage resources and to adapt aspirations over time. See Table 6.1 for a summary of the expansion and evolution of these firms via industrialized knowledge production.

OTHER ORGANIZATIONS

This section outlines findings from other organizations in the biotechnology and software industries, including Indian firms multinational corporations, and leading biotechnology and information technology research organizations and industry associations. Observations from these organizations confirms that cross-border participation in knowledge production and innovation was being driven by the use of new information and communication technologies and contributed to: (1) automating knowledge production and intellectual property generation; (2) the evolution of complexity and the creation of heteromorphic organizational form; and (3) expansion via a real options heuristics strategy.

Automating Knowledge Production Via ICTs: Generating Intellectual Property

Other biotechnology and software companies reiterated that the use of ICTs was taken for granted and that cross-border communities functioned seamlessly. Most interviewees noted that foreign firms typically outsourced non-core and some core activities to save costs and take advantage of intellectual capital available in India. One example was an Indian IT services company that diversified into providing services for the health care industry. These included such as outsourced clinical processes like radiology. Similarly, higher value-added outsourced work by another IT services company included providing business research, technology analysis, and writing patents. Thus, communication technologies that linked geographically dispersed locations seamlessly together with the higher acceptance of distributed work increased the viability of cross-border partnerships.

Most firms, including multinational organizations, focused on industrializing knowledge production via automation. Automation involved introducing global standards and norms common to the industry to gain credibility and legitimacy. All Indian software firms were ISO and/or SEI/CMM accredited, and biotechnology firms also adopted international standards and focused on obtaining certification for laboratory facilities. A director in a basic biotechnology research institution noted that international partners often examined the facilities before granting funding.

In addition, new methods were adopted for accelerating drug discovery

Table 6.1 A comparison of the expansion and evolution of four Indian biotechnology and software firms

	CodeOne	**Netcom**	**BioEnzyme**	**BioAlgorithms**
Interactions with foreign firms	Mainly as supplier and alliance partner	Mainly as supplier and alliance partner	Mainly as supplier and alliance partner	Mainly as supplier and alliance partner
Automating knowledge production via ICTs: - IT usage for communication and coordination - Adoption of global standards - IP generation - IP appropriation	- Yes; cross-border development & delivery of customized solutions - SEI/CMM level 5-certified. - Focus on building IP in the process space via accelerators (templates, models, etc) - Via patents & contracts	- Yes; offshore development of intermediate products/ customized products for overseas clients - SEI/CMM level 5-certified - Focus on building IP via products and patents - Via patents, contracts, and embedding products in a 'black box'	- Yes; provider of outsourced R&D services and contract research to overseas clients - ISO certified; labs are certified by US bodies - Focus on creating IP via products, services and patents - Via patents & contracts	- Yes; provider of contract research, bio-informatics and also biotechnology services - ISO certified; labs certified by US bodies - Focus on creating IP via bioinformatics products and contract research services - Via patents & contracts.
Heteromorphic organizational form	- Yes; multiple forms used - Hierarchies, alliances, informal and formal relationships	- Yes; multiple forms used - Hierarchies and alliances	- Yes; multiple forms used - Hierarchies, alliances, contractual relationships	- Yes; multiple forms used - Hierarchies, alliances, formal and informal relationships
Real options heuristics	- International expansion via hierarchies	- International expansion via hierarchy	- Expansion via subsidiaries in India into clinical	- International expansion via leveraging services,

Table 6.1 (continued)

	CodeOne	**Netcom**	**BioEnzyme**	**BioAlgorithms**
	- Sales offices in the U.S., Europe and Australia; development in India and China - Leveraged resources to move up the value chain from body shopping to creating IP, providing business solutions, consulting services	- Sales offices in several countries, alliances, and acquisition - IPO in 2005 - Leveraged resources and capabilities to gain leadership in the telecommunications space	research and contract research services - IPO in 2005 - Leveraged resources to build innovation capabilities & become an integrated biopharmaceutical company	partnerships and acquisitions in Europe - Leveraged resources and capabilities to acquire leadership in bioinformatics and microarrays
Current challenge	Obtaining and retaining talent for continued growth	Scaling operations	Scaling operations	Scaling operations to meet new targets

including computational methods, visualization tools, and simulation. Increasing efficiency in outsourced health care services, where adoption of technology was poor, also entailed introducing information technology:

> It's not unusual to find half a dozen data mining tools running in one department . . . one particular lab, all uniquely tailored for one particular set of scientists . . . Our real focus is how we can take your process and make it more efficient . . . And it doesn't matter what it is . . . if it's wet labs or call centers.

New artifacts such as three-dimensional models developed from CT images were created as a result of efforts to make processes more efficient. Thus, firms participating in cross-border knowledge production reduced uncertainty in the discovery and innovation process by recombining knowledge to create knowledge components or tools that could be exchanged for complementary resources.

Most firms were also aware of the need to generate intellectual property and patent innovations. For example, patenting was taken for granted in the R&D center of a multinational biopharmaceutical company, while software MNCs were just beginning to realize its importance in 2003. However, product-oriented Indian and foreign multinational companies in both industries had begun to emphasize patenting and by early 2007, most leading Indian firms focused on generating intellectual property and appropriating it via patenting. An R&D head in a biopharmaceutical company noted that their focus was on research and hence R&D spending was 30–40 percent of revenues. Most multinational firms had also established the equivalent of an R&D center in India to focus on generating intellectual property.[16] According to the head of the national IT industry association, intellectual property was typically created by gaining familiarity through servicing a particular industry that enabled the service provider to create something that was industry-specific and unique, particularly for the banking, financial services, and insurance sector. Moreover, despite the risk involved, even small and medium-sized companies had realized that focusing on innovation was a necessity because continuing to rely on generating cost efficiencies was likely to yield diminishing marginal value to customers over time.

The impetus to standardize processes to comply with global customers' needs and accelerate innovation made information technology pervasive and diffused new capabilities in every industry.[17] As noted by the head of the IT association:

> Once you globalize, when you face global competition, you have to meet global standards and efficiencies, then IT becomes a necessity for you . . . and we have seen that in industry after industry both in India and abroad, so, clearly, as the economy opens up you will get much bigger application in the use of IT in the core industry sectors.

For example, in the textile industry, as exports grew, incorporating IT became necessary to conform to EU regulations such as those regarding dyes containing banned chemicals. The only way to comply with these regulations was to track the whole supply chain.

In addition, innovation in information technology permitted new applications of ICTs in a variety of industries. Thus, in the engineering services sector, ICTs enabled engineers to experiment with designs 'in silica' instead of cutting metal, making a prototype and testing it on the shop floor. Similarly, in the area of new drug discovery, instead of synthesizing 50 000 combinations of a molecule in vitro, researchers could do the work 'in silica' to narrow down the possibilities to 100 before taking it to the laboratory. Consequently, technological innovation together with the need for IT in every industrial sector made it possible for multinational firms to outsource even high-end work related to innovation such as early stage discovery and clinical testing to accelerate drug discovery and development. Likewise, in market research, companies had graduated from doing data mining and analysis, to building models for decision-making.

Evolving Complexity: Creating a Heteromorphic Organization

Most biotechnology and software firms used a variety of organizational forms simultaneously, ranging from alliances to hierarchies. Performing contract research for multinational companies had become increasingly common, particularly in biotechnology. Moreover, most Indian biotechnology firms engaged in contract work to sustain research until they were able to develop a product internally and both informal and formal alliances were common. Collaborative work was encouraged, particularly during the early stages by funding researchers to institutions in the United States or Europe, or via alliances for specific technologies.

In established firms, research collaborations were pursued to source and explore new areas particularly via sponsored research with universities while globalization was pursued via hierarchical expansion. An executive in a leading Indian pharmaceutical firm also engaged in biotechnology noted that as companies expanded their geographical reach rapidly, they encountered the same issues as other global firms: how to manage global operations effectively. To address this problem the company had evolved a variant of a matrix structure: a geographical–functional matrix, and begun to staff the organization with people from different locations with the objective of gaining and transferring local knowledge across operations to market the company's products effectively. However, he commented that rapid expansion was facilitated by flexibility and high levels of informal communication and coordination rather than formal structure:

> I think in the beginning stages of a company, even if you look at a company that has been there for a long time but is just stepping out on the path of [internationalization] you will find this kind of rapid growth because whatever be the size of your domestic market, a foray into international markets immediately adds a quantum jump, particularly if you have done a good deal for the initial foray. So you are growing very rapidly. You know there is a long period where you kind of hibernate and try to find right partners and all of that and then you'll strike it rich and strike it right and suddenly there is this explosive growth . . . so it is not perhaps so much a formal structure, but actually one-man-driven, very deep kind of relationships within the company that make things happen . . . it is a period where irrespective of how it happened,[18] there is a lot of flexibility within the system to enable you to operate and move ahead. There is a lot of informal communication.

International collaborations were pursued even in the arena of basic research. The head of a public sector biotechnology research center noted that many scientists had been awarded grants to do specific research for global pharmaceutical companies. One reason for collaborations between Indian and foreign scientists was that they helped to validate research and improve intellectual contributions by pooling data from different countries. Research heads in various public scientific institutions noted that publication in international journals was important for advancement in leading Indian research centers and technology institutes. Thus, leading Indian science and technology institutions had also begun to participate more actively in global science and technology.

As the emphasis on collaborations suggests, using contracts and alliances as well as hierarchies where feasible was a core strategy of biotechnology firms seeking to gain access to knowledge or markets. Alliances were particularly important for start-up firms until they gained sufficient strength to participate in all stages of the technological life cycle autonomously. However, despite emphasis on collaborations and alliances, Indian firms with resources and capabilities also favored hierarchies particularly for commercialization, as expansion was perceived to depend on ownership advantage (Hymer, 1976). Similarly, foreign pharmaceutical multinationals also used alliances and contracts to explore technologies of potential future value but relied on hierarchies for commercialization to maintain their market position. Thus, a range of organizational design options was pursued depending on the firm's aspirations and its configuration of resources at any given point in time.

The pursuit of survival, growth, and expansion under uncertainty fuelled the emergence of 'heteromorphic' or H-form organization that encompasses the adoption of different structures at different stages in the firm's history to enable rapid capability-building and growth via innovation and the generation of intellectual property. This design strategy also provides the flexibility required to abandon projects that are not likely to be profitable from either a knowledge or market perspective. Thus, the design of the organization can itself be viewed as an experiment in matching strategy with structure. Much as in a game of chess, players evolve a strategy from simple rules in the course of participating in cross-border knowledge production. Examples of such rules include:

- Match quality required by best-known competitors or accreditation bodies.
- Maintain confidentiality.
- Understand where services or products fit in the innovation cycle and determine target customers.
- Adhere to contractual obligations for intellectual property generated from client-related work.
- Develop own knowledge base.
- Create new services/products.
- File own patents.
- Forge alliances.

Since different sets of activities and routines must be adopted to follow these rules, firms can easily assemble new organizational modules as required. The simultaneous adoption of such organizing principles constitutes the emergence of the heteromorphic organizations in which each module is evoked programmatically based on the configuration or state of the organization and participants at any given point in time. Repeated enactment of specific modules strengthens their performance, resulting in the creation of new capabilities and increasing complexity.

Consequently, cross-border linkages of various types at different stages of the organizational life cycle helped to evolve heteromorphic organizations, which, in turn, contributed to the development of a complex system consisting of interlinked organizations across national boundaries both at the micro level via individual collaborators, inter-firm alliances and at the national level via industry associations and research centers.

Expansion Via Real Options Heuristics

As observed earlier, organizations in each sector were able to leverage existing resources to expand and create new markets by applying real options heuristics as a strategy either implicitly or explicitly. Real options heuristics were applied in three ways.

First, local biotechnology and biopharmaceutical firms learned to leverage resources to create a market position globally by generating intellectual property derived from an insider position in the local environment. An Indian start-up biotechnology firm had invested in natural products in India because many large companies considered it too time-intensive and problematic to get involved with extraction and purification. Despite these difficulties, this company focused on developing systematic techniques for extracting pure compounds based on plants endemic in Southern India. This strategy would help the company gain differentiation advantage by producing unique products since only 10 percent of the 250 000 species of higher plants in the world had been explored for bioactivity. Other biopharmaceutical firms were focused on taking advantage of new opportunities and indigenous technology to establish niche markets by developing and launching low-cost products for tropical diseases such as malaria, tuberculosis, and hepatitis, or illnesses like diabetes that were rapidly growing more prevalent in the developing world.

MNC competitors also sought to take advantage of opportunities by using India as a base to understand a new, rapidly developing market. One biopharmaceutical MNC had launched a unit for low-cost drug discovery for diseases of the developing world and others were beginning to establish R&D centers. Incremental investments in new opportunities, exemplified pursuit of real options and were necessary for industry growth according to an executive in an Indian biopharmaceutical company who noted that: 'When more players come in, it will grow the industry. Obviously there will be a shake out. But the pie has to become bigger and this will not happen if there are too few players'.

Likewise, in the software industry, Indian software companies leveraged their strengths in producing customized software to expand by entering new areas such as medical services and a vast array of outsourced business processes. Most leading multinational software companies had also established centers for business process outsourcing, high-end services, and R&D, intensifying competition.

Second, from an aggregate industry perspective, evidence of a real options heuristics strategy is evident both in the expansion of firms and the shift from work requiring simple capabilities to more complex ones. In the IT sector, simple services were giving way to higher-end work; similarly, in the biopharmaceutical sector, contract research was growing in the discovery and development stages. Many local firms were, consequently, more willing to invest in R&D (on average about 2 percent; one biopharmaceutical firm invested 30–40 percent of revenues in R&D), provide inputs on curriculum design to educational institutions, and valued Master's degrees more highly if students' project work had industrial relevance.[19]

Third, from a societal perspective, evidence of real options heuristics can be derived from the expansion and diffusion of capabilities developed in the

IT sector into other industries and sectors of the economy. Thus, a virtuous cycle was initiated fuelled by the need to face global competition: meeting global standards and efficiencies in almost every industry necessitated the use of IT, thereby accelerating its use, and leading to greater awareness of global markets and raising global competitiveness. Besides expansion into textiles and health care for example, attempts were also being made to incorporate IT into the rural and agricultural sectors by creating new linkages. Examples include experiments like the establishment of e-platforms ('e-chaupal') in villages that allowed farmers to use technology to save 10–20 percent on input purchases and obtain 10–15 percent more for outputs.

Similarly, researchers in leading technological institutes focused on developing innovative technological solutions that were more suited to local needs (for example, products that use less power to provide the functionality required). This involved incubating technology-based start-up companies. Thus, in 1994, researchers focused on increasing the number of telephones in India from 7 million to 100 million and established a company to address the problem; they first identified the local loop (pair of copper wires) as the bottleneck in the system, and then developed a wireless local loop system. By 2005, the target of 100 million telephones in India had been surpassed: sales of the wireless local loop system were approximately US$234.37 million (INR1000 crores)[20] and the system was being used in India and 20 other countries. Likewise, in 2001, the research center also established a company to provide internet and telephone services in villages.[21] The company would provide the basic infrastructure, which included a kiosk, local language software and training at a cost of US$1166 (INR50 000) to aggregate demand and enable the local entrepreneur to run the kiosk at a profit by providing educational, health care and financial services. These services included internet-based tutorials to help students to pass English and mathematics examinations; telemedicine to help connect rural patients with remote doctors using a basic telemedicine kit, which monitors temperature, blood pressure, pulse rate, and ECG; banking via a low-cost ATM machine (at a cost of US$1166 [INR50 000] was 16 times lower than US$18 650 [INR800 000] for the traditional machine) using biometric authentication for security rather than an identity number because secrecy was not a part of village culture.

Biotechnology research centers had also incubated a few companies somewhat serendipitously, and were sometimes led by individual scientists interested in pursuing commercialization opportunities more actively.[22,23] Finally, a fledgling venture capital industry had also begun to emerge to support entrepreneurial ventures.[24]

CONCLUSION

This chapter outlined how, under uncertainty, Indian firms expanded globally by using a strategy based on real options heuristics and adopting H-form organizational design to accelerate and industrialize knowledge production.

Evidence from Indian software and biotechnology firms suggests that the use of ICTs facilitated cross-border innovation by easing communication, coordination, and codification and evolving practices to promote knowledge-sharing, thus creating the infrastructure for industrializing knowledge production. Most software and biotechnology firms learned about international markets via participation as suppliers or through forging partnerships and alliances with MNCs; many had access to expertise through founders' education, experience, and connections in the United States.[25] Moreover, access to knowledge was enhanced via technical alliances with other firms or experts in foreign (mostly US) universities and participation in industry and trade associations. Although biotechnology firms used alliances in the early stages of the technological life cycle, software firms' alliances were largely focused on downstream activities, reflecting a more mature industry and the need to integrate domain-specific business knowledge with technology. However, both software and biotechnology firms with strong capabilities and resources also used hierarchies to expand and many had established sales subsidiaries in the United States.

Thus, evidence from these firms indicates that they experimented with organizational design to accommodate the need for flexibility and access to resources and knowledge. These structures facilitated and yet controlled the rate of knowledge flow across intra-organizational units and inter-organizational units, resulting in the emergence of the heteromorphic organizational form, which comprises the simultaneous use of a range of organization forms such as short-term contracts, alliances, and hierarchies to achieve different objectives. As observed in these organizations, the presence of uncertainty in technology-intensive environments and competitive pressure to accelerate innovation created strong incentives to share knowledge across organizational boundaries. Moreover, knowledge-sharing was encouraged by aligning the interests of alliance partners. Nevertheless, knowledge-sharing in international partnerships occurs mainly in arenas where Indian firms have existing skills and when trust is secured through contracts or by assigning ownership for the knowledge generated through patents.

Patenting was common in both industries but more prevalent in biotechnology than in software; in both industries, product-oriented firms were more familiar with patenting while firms providing services perhaps relied more on customization that was difficult to replicate. However, multinational companies remained the largest producers of patents.

Indian firms also adopted a pragmatic approach and used real options heuristics to expand the scale of their operations in incremental steps by creating knowledge components that could be exchanged for complementary resources and capabilities. As Indian biotechnology and software firms gained international recognition, entrepreneurship acquired legitimacy, thus helping to create an environment that nurtures and stimulates entrepreneurial capabilities.

This analysis provides preliminary evidence of real options heuristics (ROH) as a strategy for firms in emerging economies as investments are multi-staged and incremental rather than one-shot. Moreover, using ROH as a strategy allows firms to acquire knowledge of distant technological domains and build new capabilities. It also suggests that social communities can be formed readily when there is prior knowledge of the 'codebook' by using information technology to accelerate knowledge codification and by increasing interactions with other firms via alliances and contracts, firms can extend their boundaries. A consequence of such experimentation and interactions using ROH is the evolution of markets for knowledge. Through interactions and negotiations in contracts and alliances, participants define the context, learn to communicate with one another, evolve consensus about standards in new domains of knowledge, and establish their position in the global economy. Thus, exchange is made feasible in the arena of knowledge where transactions costs are generally presumed to prohibit it (Buckley and Casson, 1976) suggesting that markets do not emerge spontaneously but are constructed.

It also indicates that national boundaries are less impermeable to the diffusion of capabilities than presumed earlier. A consequence is the emergence of firms that are global participants in new industries in regions previously inexperienced in them.

NOTES

1. The production of technological innovations and knowledge is fraught with uncertainty with respect to technology, markets, and the commercialization process.
2. This number includes other Indian firms and multinational subsidiaries of software and pharmaceutical companies.
3. These include a software association; an organization focused on facilitating technology transfer in biotechnology; a state public sector organization responsible for IT procurement, consultancy services, and training; a state-level government department – commerce and industry; and three public sector science and technology institutions.

4. A manufacturing assembly line requires an increased flow of inputs or 'throughput' to capture economies of scale and lower unit costs (Chandler, 1990). By analogy, accelerating knowledge production or innovation via an assembly system necessitates an augmented flow of knowledge inputs or components. Knowledge components are artifacts, tools, or methodologies created by recombining tacit and/or explicit knowledge.
5. Complex environments require multiple perspectives to handle complexity (Weick, 2007); analogously, strategies for such environments require complexity in organizational structure.
6. Maurice, Sellier and Silvestre (1986) found in a matched sample of French and German firms that promotion was based on demonstrated skills in the latter, while academic qualifications played a more significant role in the former.
7. Software Engineering Institute/Capability Maturity Model.
8. The term 'codebook' (Cowan et al., 2000) is used to denote deep domain knowledge.
9. While the ecological view of the firm and niches it occupies suggests a static view of the firm (Hannan and Freeman, 1977 and others), the permeability of organizational boundaries permits a more dynamic view by indicating both how firms come to occupy certain niches and how they might migrate to new ones.
10. Other works that note the importance of networks include Kogut (2000), Ghoshal (2001), and Stark (2001).
11. The case of BioEnzyme Inc. seems to support Hymer's ([1960] 1976) contention that ownership advantages are important for internationalization.
12. While historical wisdom suggests that partnerships would be eschewed when building proprietary technologies, such alliances may be used when speed in bringing new technologies to market is critical (Folta and Miller, 2002).
13. A key challenge was maintaining the entrepreneurial spirit of the company as it expanded.
14. This package was a proteomic tool that allowed firms to address several applications in the drug discovery and developmental process. The tool enables the detection of different proteins expressed by tissues using two-dimensional gel-electrophoresis. The tool helps researchers to capture, organize, and analyze protein sequence data.
15. It may also be noted that pharmaceutical multinational clients were interested in bioinformatics and other services provided by companies in India because of the cost savings and need to accelerate innovation-related processes given the long and costly development cycle for new products.
16. While Indian firms were beginning to patent, the numbers were not yet very large; multinationals (MNCs) were the largest patent producers. However, it was difficult to estimate how many patents were generated by them as most MNC patents were filed by the corporate entity in the United States or Europe (according to an interviewee at NASSCOM in 2007).
17. This is reminiscent of the contentions of Young (1928) and Rosenstein-Rodan (1943; 1944) and, more recently, others (Romer, 1986; Arthur, 1989; Murphy et al., 1989) that industrialization with new technologies that yield increasing returns can stimulate development.
18. The executive noted that it could be an individual who is 'in the hot seat', who is charismatic and has deep relationships, who gets things done.
19. Earlier, the perception in industry was that spending two years to obtain a Master's degree was a waste of time because it did not provide relevant industrial experience; this was a major discouragement to obtaining Master's degrees in technical fields and the development of research.
20. The exchange rate used for conversion: 0.02344 US dollars per Indian rupee, April 2007.
21. India has 600 000 villages and in 2001 only about one-third had a telephone connection.
22. According to the former head of a leading biotechnology research center, some scientists had also become entrepreneurs and started their own biotechnology firms.
23. Moreover, the need to promote commercialization and entrepreneurship was recognized; a leading science and technology institution was hosting a workshop on biotechnology commercialization in January 2007.

24. Some research on regional industrialization (Florida and Smith, 1993; Gompers and Lerner, 2001; Sorenson and Stuart, 2001) notes the importance of venture capital for the evolution of high-technology industries.
25. This supports Saxenian's (1994) contention that clan members form important conduits for the transfer and diffusion of knowledge.

7. From paupers to princes: the emergence of the Indian multinational corporation

As our desire is, so is our will. As our will is so are our acts. As we act, so we become.
(Brihadaranyaka Upanishad)

Based on the findings presented earlier, this concluding chapter develops a framework to explain *how* and *why* cross-border innovation occurred and how new Indian multinationals emerged and expanded overseas. The adoption of new technologies and matching organizational design transformed individual firms, and propagated changes in the macroenvironment, resulting in the creation of capabilities and a new location of excellence. In addition, the emergence of the Indian multinational ushered in sociocultural change and helped to legitimize entrepreneurship. Finally, the globalization of Indian firms also has implications for economic development.

TOWARD A FRAMEWORK FOR THE DYNAMICS OF COMPLEXITY AND EMERGENCE: ASPIRATION ADAPTATION IN CROSS-BORDER INNOVATION

Evidence from Indian firms in both 'old economy' and 'new economy' industries suggests that their interactions with foreign multinationals both as recipients of technology and suppliers of services enabled them to adapt to a changing national context and acquire capabilities necessary to participate in the global economy. Findings from various firms are used to develop a conceptual framework that maps the evolution of knowledge and capabilities and dynamics of multinational expansion.

While explanations for the existence of MNCs abound, ranging from market power and collusion (Hymer [1960] 1976; Kindleberger, 1969), to technological evolution and its life cycle (Vernon [1966] 1979; Cantwell, 1989) existing internationalization theories rarely focus on dynamics of emergence of new ventures internationally (Oviatt and McDougall, 1994). Knowledge-based perspectives rely on the social aspect of knowledge (Kogut and Zander, 1993; 1996) or transaction costs (Coase, 1939;

Williamson, 1975; Buckley and Casson, 1976) to explain international expansion. Both views lead to the conclusion that knowledge-based firms grow mainly via internalization: the former due to the embeddedness of knowledge in communities (Kogut and Zander, 1993; 1996), and the latter because of failure in the market for knowledge. Neither approach considers the impact of uncertainty on the mode of expansion. Although some studies have noted that past experience and behavior can motivate international expansion (e.g., Aharoni, 1966), most do not examine *how* international expansion of the firm co-evolves with the growth of knowledge and capabilities. Moreover, firms from emerging economies face different environmental conditions that dictate a different strategy to overcome challenges such as lack of capital, technology and skills, and a perception of poor quality. The framework suggests that technology and organization co-evolve, allowing the emergence of complex systems capable of adapting under uncertainty. It also extends behavioral theories of organization (Simon, 1965; 2000; Augier and Sarasvathy, 2003) by focusing on aspirations that fuel action and achievement (McClelland, 1959; Selten, 2001).[1]

Complex new multinational organizations from India arose as a consequence of learning from interactions with overseas partners. While technological learning was a starting point, international growth led to the evolution of organizational complexity to enable rapid adaptation under uncertainty in a complex global environment (Simon, 1956). Organizational complexity was generated by using real options heuristics (ROH) as an expansion strategy and adopting an organizational design – heteromorphic organizational form (H-form) – that permitted flexibility and adaptation in uncertain environments. Dynamic learning led to the emergence of multinationals with new capabilities in a country not regarded as an important player in world markets before the late 1990s.

This perspective on international expansion emphasizes pragmatism and experimentation in developing insights about the context that guides decision-making and action in environments characterized by uncertainty.[2] Learning and innovation were initially necessary for successful technology absorption in Indian firms, and continue to remain critical for competing through innovation in global markets as they become producers of knowledge-based components, services, and products. It also highlights the importance of focusing on *processes* as well as outcomes and suggests that, depending on environmental constraints, a variety of methods can be used to achieve organizational outcomes.

Acquiring knowledge relevant for innovation and evolving organizational complexity to cope with uncertainty depends on three critical behaviors that help learning by:

1. deciphering the context;
2. using simple rules; and
3. adapting procedures and aspirations in response to feedback.

Indian firms evolved new strategies and complex organizations and developed enhanced information-processing capabilities by adopting these behaviors in their interactions with foreign multinational firms. Over time, the interaction ritual[3] changed as they evolved, allowing for the emergence of new identities, behaviors, and a shift in power relations.[4]

Since models of emergence and innovation must incorporate learning from experience (Holland, 1998), innovation is viewed as a game or problem-solving situation requiring decisions and moves that contribute to winning. Winning the innovation game is equivalent to successful market adoption of the innovation and, potentially, higher market share for the firm.[5] The state of the game or its configuration at any point determines the play of the game from that point on and is a summary of past history for determination of all future possibilities. Thus, both initial and subsequent configurations of the game determine outcomes. At each stage, the game configuration constrains players' moves based on the rules of the game (ibid.).

Only a small number of rules are required to define a game so complicated that its possibilities cannot be exhausted. Consider a game in which there are ten possible moves from each configuration (state) including the initial configuration. If the game terminates after the two moves, there are $10^2 = 100$ ways of playing the game. If it terminates after ten moves, there are $10^{10} = 10\,000\,000\,000$ ways of playing the game. Novice players can, nevertheless, learn to outdo experts despite the difficulty of predicting outcomes from the large number of combinations of game configurations (ibid.).

Making successful moves in games dictates assessing other players' likely responses. When a game is played repeatedly, unknown aspects of other players' strategies are revealed. By observing other participants, players can learn what they are likely to do when confronted with decisions at different choice points, thus enabling them to construct a model of their opponent's strategy. However, a strategy cannot be defined by listing all the game states and possible moves because there are too many states. Hence, strategies can be defined via a set of rules or heuristics.[6] In chess, these rules embody principles like 'build a strong pawn formation' or 'control the center'. Such rules focus on features that occur frequently and are relevant to decisions at various points in the game. Thus, states with similar features can be reduced to clusters suggesting similar decisions and moves. Repeated plays of the game can be used to discover and combine building blocks (e.g., heuristics, features) to construct a feasible strategy (ibid.).

By analogy, participating in the innovation game reveals knowledge about other players' strategies. Through repeated interactions participants learn to anticipate one another's moves, and thus enhance their repertoire of available strategies. Since speed is of the essence, contextual knowledge is critical for operating in uncertain environments. Players gain familiarity with the context and knowledge about other players by participating in the game using simple rules and strategies. Even though the overall system is fully defined as when a computer is supplied with the rules of the game, prediction is difficult even after extended observation. Strategies co-evolve, each strategy adjusting to its experience with its opponents. This co-evolution exhibits the creativity inherent in evolutionary processes.

Indian firms used apprenticeship to understand the features of the game through attempts to 'match' lead firms' moves.[7] They also experimented with rules and moves like 'acquire technology', 'establish alliances', and 'continuously improve quality' depending on the resources and capabilities available. Making these moves, in turn, yielded more information, resulting in new rules such as 'match quality of US MNCs', 'differentiate product offerings', 'export to other developing countries', 'establish sales offices and/or subsidiaries overseas', and 'use a variety of organizational forms'.

Moreover, aspirations were adjusted as learning occurred. Though goals were initially focused primarily on learning-specific tasks, such as acquiring technical knowledge, as these tasks were mastered, a new identity emerged, initiating a self-fulfilling prophecy of success. Firms evolved from 'bricoleurs' (Levi-Strauss [1962] 1966) using craft-based methods of knowledge production to 'engineers' using industrial technologies to manufacture knowledge and leverage innovations successfully to expand the firm.[8] New aspirations also arose from interacting and identifying with leading firms, suggesting an evolution towards a 'geocentric' organization (Permutter, 1969[9]). Prominent among these was the desire to be ranked as global leaders. Consequently, attempts to match leading global players led to the adoption of new attitudes, orientations, and rules that accentuated the importance of entrepreneurialism, opportunity-seeking, and international expansion[10] (see Figure 7.1 for an outline of the dynamics of cross-border innovation).

In conclusion, evidence from the studies outlined in earlier chapters suggests that Indian firms used a staged approach to globalization during 1993–2003, evolving new strategies and moves as they deciphered the changing context. As noted earlier in the book, a staged approach is used

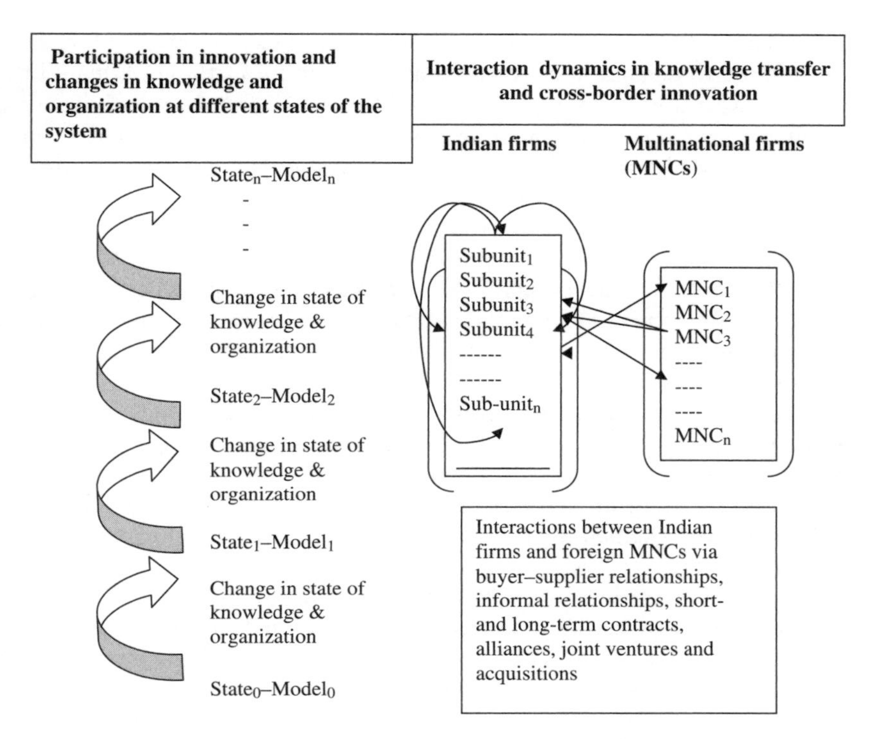

Figure 7.1 Evolving complexity: a model of the dynamics of knowledge transfer and cross-border innovation

as a device to denote demarcations in capabilities rather than a fixed duration and does not imply that evolution is necessarily linear or that progress is inevitable; thus, it does not rule out the possibility that some firms may evolve faster than others while others may remain at a particular stage if conditions for learning and evolution are not conducive.

In the first stage, firms focused on building a community of practice; most firms (except Bearings) forged strong relationships with their technology suppliers and created an organizational environment that enhanced intra- and inter-organizational communication and coordination to facilitate the diffusion of learning via apprenticeship. Earthmovers relied on technology collaborations to diffuse technical knowledge of hydraulic machines and social and organizational know-how required to convert the organization from a sequential to a concurrent manufacturing system. As capabilities were developed, reliance on foreign suppliers was reduced.

Similarly, Steelworks relied on direct imports of technology and foreign consultants to create a local community of practice in India. A team of internal experts was developed to install the new technology and diffuse learning to the rest of the organization. Here too, the group focused on sharing knowledge and collaborating intensively to enhance the organization's capacity to absorb knowledge rapidly.

Unlike Steelworks and Earthmovers, Bearings foundered initially because of differences in the priorities of joint venture partners and impediments to internal collaboration arising from the presence of expatriates and differences in work style and culture. However, when emphasis on collaboration was restored, the Indian JV experienced performance improvements.

Other Indian manufacturing firms also replicated similar local communities using their relationships with foreign technology suppliers and joint venture partners. Consequently, the first stage of evolution involved creating conditions in the domestic environment to foster the absorption of new technologies by augmenting organizational coordination and communication capabilities.

The second stage involved enhancing the ability to absorb and process knowledge more efficiently and speedily by institutionalizing learning across the firm via best practices. A preoccupation with doing everything internally gave way to a focus on external search and benchmarking quality against global manufacturers. This stage was transitional; executives realized that not only new technology, but also ideas about organization could be adopted from external sources. Consequently, efforts to strengthen organizational information-processing capacity and match global quality in manufacturing were accelerated. All three organizations reinforced these priorities by introducing benchmarking and other best practices. For example, R&D engineers in Earthmovers noted that design capabilities had evolved by following new methods; similarly, productivity and quality improved by installing self-monitoring and problem-solving principles on the shop floor. In addition, the discourse evolved to incorporate these concepts.

After enhancing internal communication and coordination, similar quality and productivity improvements were found in Bearings. Expatriate managers began to learn about domestic resources and developed local suppliers to reduce costs and reliance on technology imports. Likewise, in Steelworks, practices adopted in the new blast furnace were institutionalized and diffused across the organization via modernization projects – top management professed that the goal was to become a 'learning organization'.[11]

The third stage was marked by the adoption of new technologies and the internet to facilitate Indian firms' participation in global markets as suppliers of products and services. During this stage, firms attempted to integrate ideas from both internal and external sources to accelerate innovation. Moreover, collaborative technologies eased cross-border knowledge production since knowledge itself was increasingly embedded in artifacts reliant on information technology. Industrializing knowledge production led firms to use a variety of organizational mechanisms over the course of the technological life cycle[12] and adopt modular organizational designs. This resulted in the emergence of information-based 'heteromorphic' organizations, a form that encompasses alliances, hierarchies, and a variety of contractual arrangements to enable firms to create and operate a portfolio of real options. This form also allows the firm to expand via 'real options heuristics', while retaining the flexibility to exit if pursuing a particular option is not viable. Consequently, geographical borders are rendered less relevant for knowledge production.[13]

By the third stage, firms in all three sectors – manufacturing, software, and biotechnology – had embraced information technology to enhance collaboration and communication greatly with external partners both in terms of frequency and regularity, in contrast to the sporadic levels of communication possible earlier. Moreover, the use of technology eased the creation of communities of practice to the extent that cross-border collaboration was taken for granted. Earthmovers pursued other technology partnerships for indigenously designed excavators although the Japanese JV partner had increased its stake. Engineers noted that detailed design collaboration was possible since drawings and other documents could be readily transmitted over the internet; the only requirement was that both parties use standardized software. Thus, many projects ranging from design to manufacturing were pursued simultaneously, subject to financial and managerial constraints.

At this stage, Bearings was viewed as an established Indian subsidiary and a stepping stone to gain access to R&D and design resources. Consequently, attempts were made to align the Indian manufacturing plant with the priorities of the global manufacturing center and use the Indian experience to replicate a manufacturing center in China. In addition, the Indian subsidiary established an R&D center in Bangalore. Manufacturing was standardized to make the Indian subsidiary a specialized and seamlessly integrated unit in the multinational network. The role of the Indian R&D center was to enable the multinational to standardize and accelerate innovation. These moves are evidence of the use of real options heuristics in decision-making.

Steelworks also engaged in organizational reconfiguration to accommodate changes in strategy and augment internal capabilities. New rules were learned and adopted by interacting with partners and consultants.

These included making strategic planning a continuous activity, using mechanisms like the balanced scorecard to align corporate and functional strategies, focusing on acquisitions, and investing in new e-businesses. The combination of IT and organizational reconfiguration to coordinate and manage tasks facilitated knowledge decomposition and recombination and enabled the peripheral activities to be outsourced. These moves led to perceiving the organization as a portfolio of businesses; decisions about whether to pursue a business depended on aspirations, resources, market conditions, and competition. In addition, they also set the organization on a trajectory of international expansion.

Evidence from the software and biotechnology firms suggests that they were at the third stage by 2003. Indian software firms took full advantage of new communication technologies and were already recognized as international players. Biotechnology firms, although recently established, also took IT-based cross-border communication with peers for granted. Moreover, knowledge production in these industries had become increasingly reliant on the digitization of knowledge. The ability to share knowledge instantaneously via ICTs, together with the desire of foreign multinationals to save costs by accessing cheaper resources overseas, propelled Indian firms into cross-border knowledge production. In both industries trust was evolved by adopting international quality standards for technology, facilities, and practices. Interactions between Indian and foreign firms led to the understanding that contractual obligations must be fulfilled to maintain credibility. Consequently, trust was more widely diffused, allowing players to participate more freely in cross-border collaborations. Particularly in the biotechnology industry, a wide range of organizational mechanisms was used to gain access to new knowledge. These ranged from informal and formal partnerships with Indian and foreign universities for discovery, to strategic alliances for development and commercialization.

Although earlier research suggested that proprietary knowledge was usually appropriated by establishing hierarchies (Buckley and Casson, 1976; Teece, 1977), evidence from 'new economy' firms indicates that biotechnology firms evolved ways to share the benefits of joint research by relying on contracts and patents to establish trust, corroborating Lewis's (1970) seminal insight that the transformation from 'status' to 'contract' in developing economies is profound.

Entrepreneurial firms lacking resources exploit alliances and joint ventures to gain access to new knowledge, acquire capabilities, and expand internationally (Oliver, 1994; Powell et al., 1996) while hierarchies are used by large firms. Consequently, hierarchies are not the only means of expansion. However, this does not suggest that hierarchies are likely to be abandoned in favor of other forms. The evidence suggests 'heteromorphic' organizational forms evolve as both entrepreneurial and established firms make use of a wide variety of forms concurrently and over the technological life cycle.

While biotechnology firms rely on alliances and other formal and contractual partnerships early in the technological life cycle, software firms are more likely to engage in downstream partnerships for expertise in specific businesses to aid commercialization. One reason may be that the biotechnology industry in India is at an earlier stage of development than the relatively more mature software industry.

In summary, at each stage, most firms followed a similar experimental process that culminated in step changes in strategy, rules, and organization. Each stage was also accompanied by a reconfiguration of the environment resulting from changes in government policy with respect to privatization and the ownership rights of foreign MNCs. By industrializing knowledge production, Indian firms established their role as suppliers of knowledge-based services and components to foreign MNCs and contributed to defining their niche in globally distributed innovation chains.[14] Attempts by Indian firms to differentiate themselves in the global economy also led to increased specialization and the creation of a new location of knowledge production. Their participation in global innovation and that of MNCs seeking to take advantage of India as a new market and as a source of knowledge resources, as evidenced by business press reports in the United States indicates that Indian firms are on the threshold of global expansion.

EXCEPTIONS TO THE PATTERN

Although this framework is developed from close examination of a small number of firms, the data are obtained from 'focal' firms in various manufacturing industries (ten firms) and two knowledge-intensive industries (18 organizations).[15] Two exceptions to the pattern were found: Bearings and a multinational pharmaceutical firm with operations in India. However, both firms altered their strategies and behavior in accordance with the framework presented here. In the case of Bearings, when adjustments were made in 1993–94 to create a local community of practice in India, the firm began to experience rapid improvements in productivity and quality. The Indian subsidiary of the pharmaceutical MNC (which had been operating in India since the 1950s) was being aligned with global manufacturing operations in 2003. R&D was not a major focus of the Indian subsidiary until later, although a center for clinical trials was established in Bangalore and R&D operations outside Mumbai to cater to unmet medical needs and exploit local commercial opportunities. However, the entry of other multinationals in India and the availability of resources for R&D in India helped shift the strategy to introducing drugs suitable for the local market and researching and testing new drugs for diseases prevalent in developing countries.

IMPLICATIONS FOR DEVELOPMENT

These findings have implications for the development of emerging economies. The wider diffusion of standards and best practices, adoption of ICTs in every industry, and development of technological and entrepreneurial skills in new industries suggests that development can be jump-started by investing in building capabilities in sunrise industries as noted by Nelson and Pack (1999). These capabilities trigger a virtuous cycle of learning, innovation, expansion, and cultural transformation (Arthur, 1989; Murphy et al., 1989; Stiglitz, 2004) in emerging economies.

As evidenced by the evolution of Indian firms these capabilities are not built in a vacuum, but in the context of learning via technology acquisition. Hence, establishing rules to stimulate investment in new technology and learning can help initiate development. This research also suggests that change can be implemented successfully by adopting modes of organization to match new technology at each stage of evolution, enabling firms to reallocate and use resources more effectively. Reorganization is especially critical in emerging economies with scarce resources and skills. The challenge of matching multinational competitors spurred a widespread response among Indian firms to improve quality by benchmarking global leaders. Likewise, incentives to do so must be in place in developing countries to overcome the liability of a perception of low quality.

ALTERNATIVE EXPLANATIONS

While many explanations for international expansion exist, these are generally derived from studies of the expansion of US or European multinational enterprises; few researchers have focused on the internationalization of firms in developing countries. These explanations for the existence and emergence of MNCs include market power (Hymer [1960] 1976), oligopolistic rivalry (Knickerbocker, 1973), the product life cycle of technological innovation (Vernon [1966] 1979), internalization of markets across borders as a result of failure in the market for knowledge (Buckley and Casson, 1976; Casson, 1997), or the need to replicate social communities for knowledge transfer (Kogut and Zander, 1993). These explanations emphasize that hierarchies are the mechanism of choice for overseas expansion since the chief aim is to transmit and exploit proprietary knowledge developed in industrialized countries. While Vernon's ([1966] 1979) revised work on the technological life cycle does suggest that innovations can originate in developing countries, as does Cantwell (1989) in his evolutionary theory of the multinational, this idea is not explored in depth. In addition, the dynamics of *how* new firms emerge from developing countries and the processes by which they expand internationally in the face of competition from incumbent global multinationals are not well understood.

Moreover, the logic explaining the internationalization of US and European firms is not necessarily the same as that for firms from developing countries. The reason for this is that historical circumstances were different; earlier perspectives on internationalization were developed during a time when MNCs from industrialized countries were dominant players in oligopolistic international markets. Also, past theories did not take into account how uncertainty and new information technologies that simplify codification enable international expansion via the adoption of a new strategy and structure – ROH strategy and H-form organization.

However, evidence from the experience of other developing countries suggests support for the perspective developed in this book by situating the phenomenon under observation in India in a wider context.[16] For example, there are similarities with the emergence and global expansion of Korean firms. Like leading Korean firms, Indian firms emphasized capability-building and knowledge diffusion within the organization and the industrial community. Korean companies also evolved new capabilities by mastering 'old economy' industries such as steel and textiles before venturing into new industrial sectors (Enos and Park, 1988). This framework is also corroborated by the success of specific industries such as the aircraft industry in Brazil. Aircraft manufacturing capabilities were evolved using similar methods by EMBRAER (Empresa Braziliera de Aeronáutica, SA), founded in 1960 in Brazil, and now a leading manufacturer of commercial and defense aircraft (Nelson, 1993). Similar evolutionary processes underlie the early history of industrialized countries like Britain and the United States (Licht, 1995). The early history of Japanese industrial evolution suggests that capabilities evolved similarly from manufacturing to innovation in the automobile industry (see Pascale, 1984 for details on the evolution of Honda; Nelson, 1960; 1993).

Other research on Japanese and Korean multinationals that rose to challenge US multinationals earlier (Dore, 1973; Mansfield, 1988; Dore and

Sako, 1989; Aoki, 1990) suggests that capability-building is a key strategy in the rise of firms from a new location and that firms' progression towards internationalization occurs in a step-wise, non-linear fashion. Examples include capability development in manufacturing steel in Korea; some aspects of POSCO's (Pohang Iron and Steel Co.) evolution from a government-initiated domestic enterprise to a private global corporation are reminiscent of development of the steel industry in India (Lall, 1987; Enos and Park, 1988; Nelson, 1993) including the pursuit of global aspirations (Ungson, Steers, and Park, 1997). In documenting the Spanish experience, Guillen (2003) emphasizes the support of local institutions and indicates a similar process of capability-building. At the level of the firm, this framework is supported by evidence that balancing vertical integration and strategic outsourcing improves firm performance and product success (Rothearmel, Hitt, and Jobe, 2006) while Metiu's (2006) study of distributed teams in India and the United States indicates the possibility of originating innovation from India.

Yet, while the Indian experience is similar to Korea's or Brazil's in terms of capability-building, it is unique in terms of the sheer size of the Indian economy and its potential impact on the world economy. Also, Korea began its modernization efforts earlier, in a different historical context. In contrast, India continued to pursue import substitution until the 1980s. Consequently, Indian firms were largely oriented toward the domestic economy and began to focus externally only when the Indian market was opened to foreign competition.

A comparison with the Chinese experience reveals sectoral differences in each country's trajectory of capability-building suggesting that strong capabilities can emerge in different sectors and that there are several paths to internationalization. China differs in its emphasis on manufacturing prowess, export-led growth, and resulting trade surplus with the United States of US$13 billion (*NY Times*, 13 June 2006). Yet, it is similar with regard to aspirations for global expansion as recent attempts by Chinese organizations to acquire US businesses suggest.

Differences in the globalization trajectory of Indian firms stem from context: the rise of Indian firms coincided with that of the internet, a disruptive technology that created the potential for new forms of organization to emerge. It was also a consequence of earlier investments in building skills. This study provides preliminary evidence that information and communication technologies allowed new Indian players to participate in worldwide markets. Moreover, the concurrent use of various organizational forms such as alliances, hierarchies, and other informal mechanisms to speed the acquisition of knowledge and capability-building in India suggests that the expansion of the firm across borders need not be a consequence of failure in the market for knowledge. Rather, it can be viewed as a result of deliberate experimentation by firms to acquire capabilities and participate as members of an international community.

LIMITATIONS

The framework developed in this book attempts to capture a complex phenomenon that is, as yet, at an early stage and requires further exploration. The question remains as whether a similar expansion process would be stimulated in the absence of uncertainty. A second limitation is that because the focus was on the dynamics of international expansion at the level of the firm and industry, macro factors such as exchange rate changes, foreign direct investment (FDI) flows, national culture, and institutions were not examined.[17] This book presents a potential strategy and coordination mechanisms that might be explored. While these appear to be deliberate, the evidence from the current study is not conclusive, and needs to be tested more formally using other methods such as simulation (Davis et al., 2007) and large sample longitudinal studies. Our study also suggests that the adoption of ICTs and global standards is correlated with the adoption of real options heuristics strategy and heteromorphic organizational form. Further investigation is needed to determine whether and to what extent these variables can vary independently.

CONCLUSION

This study of the globalization of Indian firms focuses attention on processes and highlights the importance of aspirations in building new capabilities and complex organizations. It also suggests that innovation is a consequence of adaptation and adjustment: even when firms engage in imitation, an altered context dictates interpreting knowledge differently and adjustments to suit the new environment. Learning occurs at each stage; increased interaction with foreign firms accelerates learning and results in changes in structure, cognition, and behavior. Indian firms experience an identity shift aligning them with their overseas partners. A conclusion is that meta-change requires mastery of the identity-shifting process.

Although lack of financial and other resources are important constraints in determining the trajectory of expansion in emerging economies, it is access to knowledge and capabilities that plays a decisive role in determining the growth and international expansion of firms in emerging economies. Therefore, while reinforcing the idea that differing initial conditions are likely to influence the trajectories of firms originating in different countries, this examination of international expansion also emphasizes that imitation leads to similarity in aspirations and motivations and a desire to gain recognition as valued members of a global community. The notion that cognitive collocation can be substituted for geographic collocation in certain instances, especially in knowledge-intensive industries, suggests that the debate about convergence is moot, at least in sectors where countries have strong capabilities. Moreover, cross-border interactions between firms in different geographic locations require alterations in strategy for all participants, suggesting a shift in focus to co-evolution within networks of relationships.

Also, given that firms use a variety of organizational mechanisms for international expansion and that cross-border partnerships are used even for knowledge production and innovation, the boundaries of firms and nations are not necessarily impermeable (Kogut, 1991).

The adoption of market-based values and orientations particularly in the domain of knowledge, is not without problems. While the euphoria generated by entrepreneurial success hastens the diffusion of attitudes associated with industrialized production, it risks the loss of diversity in orientations and values associated with craft-based production. Industrialized production has been argued to hasten deskilling through the codification of tacit knowledge and displace 'craft-based' methods of knowledge production even in industrialized nations.[18] The irony is that codifying tacit knowledge diffuses expertise and renders the expert unnecessary. Similarly, embedding experts' knowledge into computerized systems to make knowledge production more efficient reduces the need to rely on individual experts. Such transformations and displacements highlight issues of power and control embedded in systems of production.[19]

An implication is that it is important to ease the transition from craft to industrialized systems by providing training for those undergoing displacement. Likewise, knowledge workers must take responsibility for continually enhancing their own skills and capabilities.

Ownership issues are also surfaced when knowledge production is industrialized. The ability to codify and transform knowledge into tradable intellectual property has significant implications. In such a situation, knowledge workers like software engineers, biologists, or experts with the requisite specialized skills, are the new 'princes'. In contrast, participation in the benefits of globalization may be increasingly difficult for marginalized individuals and groups. Lack of participation by such groups may be detrimental as traditional knowledge embedded in 'craft-based' systems that are not industrialized is endangered. In addition, over-emphasis on 'marketable' capabilities may create a short-term orientation, resulting in the neglect of future capabilities. At risk also is the sustainability of rapid industrialization as entrepreneurial aspirations are unleashed. The acceptance of a unitary mode of participation that creates distinct winners and losers is also problematic. These implications emphasize the need for building capabilities and managing aspirations of those unable to participate in knowledge work.

Industrializing knowledge production substitutes the 'market' for the 'bazaar' in the domain of knowledge, changes the norms and rules of interaction from those of a 'community' to those of a 'market',[20] and constitutes a profound shift in orientation from status to contract (Lewis, 1970). It also surfaces issues such as increased private sector participation in knowledge creation and appropriation via patenting, thus accelerating the privatization of knowledge that was previously in the public domain. Consequently, policies must be formulated to maintain an intellectual commons for future generations. Continued government funding of basic research and participation by knowledge workers in 'open innovation' can ensure that contributions of knowledge to the public domain are maintained. Another difficulty highlighted by the success of industrialized knowledge work in specific industries is the potential neglect of others to the detriment of the entire ecology.

Finally, this research also suggests that the boundaries between firms and markets are not fixed but negotiated in inter-firm interactions. As new forms of participation evolve via different modes of organization and a different division of labor, new norms, values, and identities are also likely to emerge, suggesting that the moral basis of human functioning is embedded in our capacity for social and organizational innovation (Durkheim [1933] 1984). Moreover, greater integration of emerging economies with the global economy also calls for the development of appropriate social architectures to facilitate the emergence of a 'global civilization' through dynamic syncretism of diverse values and ideologies (Permutter, 1991).

FUTURE DIRECTIONS

The emergence of the Indian multinational is a phenomenon that has only recently gained visibility and needs to be studied further. Future research could examine the impact of the Indian multinational on global competition, outward flows of FDI (foreign direct investment) from India, and cross-border merger and acquisition activity by Indian firms.

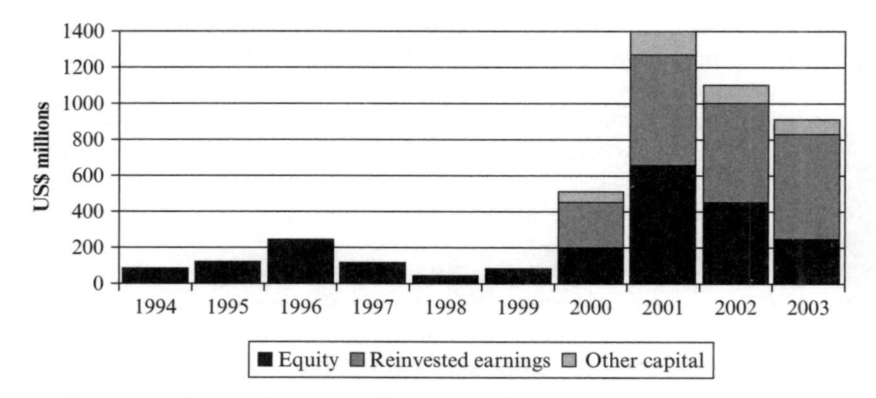

Note: Data for total FDI flows abroad from 2000 include equity, intra-company loans, and reinvested earnings. Data before 1994 are not available.

Source: UNCTAD (www.unctad.org/fdistatistics), based on Reserve Bank of India.

Figure 7.2 Outward foreign direct investment from India (1994–2003)

Heightened competition, particularly in mature industries, is likely to lead to greater merger and acquisition activity as India becomes an increasingly important location. As we have seen, although the globalization of Indian firms is at a nascent stage, expansion strategies have already begun to include mergers and acquisitions. Consequently, both inward and outward foreign direct investment (FDI) flows are likely to rise (see Figure 7.2 and Tables 7.1–7.5 for details on outward FDI flows, merger and acquisition activity, and the most attractive global business locations).

The entry of multinationals, while creating opportunities for employment in new industries has led to concomitant changes in local mores and the social environment. Cultural change, while notoriously difficult, has been initiated as a by-product of adopting work practices of global organizations; thus, new norms of individualism and achievement common in the United States are more prevalent, particularly in urban areas. Consequently, it would be fruitful to study the expansion and development trajectory of US, European, Japanese, Korean, and other multinationals in India and examine the impact of their interactions with Indian firms on innovation, entrepreneurship, sociocultural and institutional change.

Table 7.1 Geographical distribution of approved Indian outward FDI flows, fiscal years 1996–2003 (millions of dollars; percentage)

	Fiscal year					Total	
Economy	1996–2000	2000–01	2001–02	2002–03	2003–04*	FY 1996–03	Share
United States	378.5	734.2	428.1	185.3	138.7	1864.8	18.8
Russian Federation	3.3	3.5	1741.9	0.2	..	1748.8	17.6
Mauritius	221.6	242.3	154.5	133.4	160.9	912.6	9.2
Sudan	..	..	..	750.0	162.0	912.0	9.2
British Virgin Islands	752.1	18.0	6.4	3.3	2.2	782.0	7.9
United Kingdom	269.8	55.3	85.5	34.5	98.1	543.2	5.5
Hong Kong, China	391.4	37.6	16.1	14.8	13.1	473.1	4.8
Bermuda	156.9	0.7	75.0	29.0	14.7	276.3	2.8
Vietnam	0.4	0.2	228.2	0.1	0.0	228.9	2.3
Singapore	88.5	39.4	25.0	46.8	13.5	213.2	2.1
Oman	139.8	64.9	0.2	0.4	1.5	206.7	2.1
Netherlands	49.1	65.7	43.1	15.9	29.9	203.7	2.1
United Arab Emirates	87.2	11.3	11.8	12.6	29.5	152.3	1.5
Australia	2.6	2.5	1.9	95.0	41.3	143.3	1.4
Iran	59.2	..	..	43.6	0.1	102.9	1.0
China	17.1	7.9	13.3	29.6	19.8	87.7	0.9
Kazakhstan	3.2	..	1.3	0.1	75.0	79.5	0.8
Nepal	45.5	10.9	10.6	5.7	5.1	77.8	0.8
Austria	26.3	0.5	50.9	..	..	77.6	0.8
Sri Lanka	51.8	8.4	1.4	6.6	6.0	74.2	0.7
Malta	..	..	21.7	24.4	21.0	67.0	0.7
Ireland	31.8	0.2	11.4	..	0.0	43.5	0.4

Table 7.1 (continued)

Economy	Fiscal year					Total	
	1996–2000	2000–01	2001–02	2002–03	2003–04*	FY 1996–03	Share
Italy	11.7	30.5	0.0	0.1	0.0	42.3	0.4
Malaysia	33.1	4.8	1.4	0.8	1.4	41.5	0.4
Thailand	22.1	0.4	2.6	7.7	7.3	40.2	0.4
Indonesia	7.8	..	12.4	0.1	19.3	39.6	0.4
Morocco	32.5	..	..	..	..	32.5	0.3
Libya	..	..	30.0	..	..	30.0	0.3
Others	255.7	42.9	52.4	32.5	45.7	428.5	4.3
Investment in all countries	3138.9	1382.2	3027.0	1472.2	906.1	9925.6	100.0
Memorandum:							
Developed countries	809.0	913.3	651.5	367.6	346.2	3086.9	31.1
Developing Countries	2329.9	468.8	2375.5	1104.6	560.0	6838.7	68.9

Note: Data consists of equity, loan, and guarantee.

Source: UNCTAD (www.unctad.ord/fdstatistics), based on Ministry of Finance, India.
* Covers April–December 2003.

Table 7.2 Industry distribution of cross-border M&A purchases by Indian companies 1996–2003 (number of deals)

Sector/Industry	1996	1997	1998	1999	2000	2001	2002	2003	1996–1999	2000–2003
Total industry	8	13	13	26	55	35	35	57	60	182
Primary	0	1	0	2	0	0	1	2	3	3
Secondary	6	5	7	17	22	14	15	30	35	81
of which:										
Food, beverages, and tobacco	1	0	1	0	2	3	1	3	2	9
Textile, clothing, and leather	0	0	0	2	0	0	1	0	2	1
Printing, publishing, and allied services	0	0	0	0	0	0	1	0	0	1
Oil and gas; petroleum refining	0	0	0	1	0	3	1	3	1	7
Chemicals and chemical products	4	3	6	5	15	3	7	11	18	36
Rubber and miscellaneous plastic products	0	0	0	0	1	0	0	0	0	1
Stone, clay, glass, and concrete products	0	0	0	1	0	1	0	1	1	2
Metal and metal products	0	1	0	0	1	2	0	4	1	7
Machinery	0	0	0	0	0	0	3	2	0	5
Electrical and electronic equipment	1	1	0	6	2	2	0	5	8	9
Motor vehicles and other transport equipment	0	0	0	1	0	0	1	1	1	2
Measuring, medical, photo equipment, clocks	0	0	0	1	1	0	0	0	1	1
Services	2	7	6	7	33	21	19	25	22	98
of which:										
Electric, gas, and water distribution	0	0	0	2	0	1	0	0	2	1
Construction firms	0	0	1	0	1	0	0	0	1	1
Trade	0	0	1	0	2	1	0	0	1	3
Transport, storage, and communications	0	1	1	1	1	0	3	3	3	7
Finance	1	1	1	2	9	3	5	3	5	20
Business activities	1	5	2	2	20	16	11	19	10	66

Table 7.2 (continued)

Sector/Industry	1996	1997	1998	1999	2000	2001	2002	2003	1996–1999	2000–2003
of which:										
Prepackaged software	0	0	0	1	10	4	1	9	1	24
Real estate mortgage bankers, and brokers	0	0	0	1	0	0	0	0	1	0
Business services	1	4	2	0	8	12	10	10	7	40
Advertising services	0	1	0	0	2	0	0	0	1	2

Source: UNCTAD, cross-border M&A database.

Table 7.3 Geographical distribution of cross-border M&A purchases by Indian companies 1996–2003 (number of deals)

	1996	1997	1998	1999	2000	2001	2002	2003	1996–99	2000–03
TOTAL WORLD	8	13	13	26	55	35	35	57	60	182
Developed economy	3	7	3	8	24	16	18	22	21	80
Western Europe	2	3	2	3	6	4	11	10	10	31
European Union	2	3	2	3	6	4	11	10	10	31
Denmark	–	–	1	–	–	–	–	–	1	–
France	–	–	–	–	–	–	–	1	–	1
Germany	–	–	–	–	2	2	2	1	–	7
Ireland	1	–	–	–	–	–	–	–	1	–
Italy	–	–	–	2	–	–	–	1	2	1
Portugal	–	–	–	–	–	–	1	–	–	1
United Kingdom	1	3	1	1	4	2	8	7	6	21
North America	–	1	1	4	15	11	7	10	6	43
United States	–	1	1	4	15	11	7	10	6	43
Other Developed Countries	1	3	–	1	3	1	–	2	5	6
Australia	1	2	–	–	3	1	–	2	3	6
New Zealand	–	1	–	–	–	–	–	–	1	–
South Africa	–	–	–	1	–	–	–	–	1	–
Developing economy	5	6	10	18	29	18	15	35	39	97
Africa	–	1	1	–	–	–	–	1	2	1
Sudan	–	–	–	–	–	–	–	1	–	1
Zambia	–	1	1	–	–	–	–	–	2	–
Latin America and the Caribbean	–	1	–	–	1	–	–	1	1	2
Bermuda	–	1	–	–	–	–	–	1	1	1
Cayman Islands	–	–	–	–	1	–	–	–	–	1
Asia	5	4	9	18	28	18	15	31	36	92

Table 7.3 (continued)

	1996	1997	1998	1999	2000	2001	2002	2003	1996–99	2000–03
West Asia	–	–	–	1	–	2	–	2	1	4
Oman	–	–	–	1	–	–	–	1	1	1
United Arab Emirates	–	–	–	–	–	2	–	1	–	3
South, East, and Southeast Asia	5	4	9	17	28	16	15	29	35	88
China	–	–	–	–	–	–	–	2	–	2
Hong Kong, China	–	–	–	–	1	–	–	–	–	1
India[a]	5	4	8	11	23	14	13	21	28	71
Indonesia	–	–	–	–	2	–	–	–	–	2
Malaysia	–	–	1	–	–	1	–	1	1	2
Myanmar	–	–	–	–	–	–	1	–	–	1
Nepal	–	–	–	–	1	–	–	–	–	1
Philippines	–	–	–	–	–	–	–	1	–	1
Republic of Korea	–	–	–	–	–	–	–	1	–	1
Singapore	–	–	–	3	1	1	1	1	3	4
Sri Lanka	–	–	–	3	–	–	–	2	3	2
The Pacific	–	–	–	–	–	–	–	2	–	2
Fiji	–	–	–	–	–	–	–	2	–	2
Central and Eastern Europe	–	–	–	–	2	1	2	–	–	5
Czech Republic	–	–	–	–	–	–	1	–	–	1
Hungary	–	–	–	–	1	–	–	–	–	1
Poland	–	–	–	–	–	–	1	–	–	1
Romania	–	–	–	–	1	–	–	–	–	1
Russian Federation	–	–	–	–	–	1	–	–	–	1

Note: [a] Refers to foreign affiliates in India.

Source: UNCTAD, cross-border M&A database.

Table 7.4 Representative acquisitions by Indian firms (post-2000; US$)

Company	Deal Value (US $ millions)
Tata Steel–Corus	12,100.0
Hindalco–Novelis	6,000.0
Ranbaxy–Merck Generics	6,000.0
Suzlon Energy–REPower	1,200.0
ONGC–Greater Nile Oil Project	766.1
Dr Reddy's–Betapharm Arzneimittel GmbH	570.0
VSNL–Teleglobe International Holdings	254.3
Tata–Tetley Tea	431.0
Taj–Ritz-Carlton Boston	170.0
Tata Nat Steel–Vietnam Inc.	41.0

Source: Indiapost.com, 3 April 2007.

Table 7.5 Most attractive global business locations: responses of experts and transnational corporations (TNCs)

Responses from Experts		Responses from TNCs[a]	
Country	%	Country	%
1. China	85	1. China	87
2. United States	55	2. India	51
3. India	42	3. United States	51
4. Brazil	24	4. Russian Federation	33
5. Russian Federation	21	5. Brazil	20
6. United Kingdom	21	6. Mexico	16
7. Germany	12	7. Germany	13
8. Poland	9	8. United Kingdom	13
9. Singapore	9	9. Thailand	11
10. Ukraine	9	10. Canada	7

Note: [a] Countries are ranked according to the number of responses that rated each as the most attractive location.

Source: UNCTAD (www.unctad.org/fdiprospects), World Investment Report 2005.

NOTES

1. While Hamel and Prahalad (1989) and Prahalad and Hamel (1990) had earlier noted the importance of stretch targets, they did not discuss the dynamics of international expansion.

2. Pragmatism is used in the sense of Dewey (1988) and others, not in the sense of encouraging relativism. The emphasis on practice does not deny the existence of ideals or an absolute ethics; Rorty (1961) notes that although pragmatists like Pierce emphasize practice rather than idealized conceptions, they do so because they wish to avoid resorting to a mechanistic understanding of things by taking the context into account. In the case of language, Pierce's view is similar to that of the later Wittgenstein who argued that all forms of reductionism generate infinite regress.
3. The term 'interaction ritual' is taken from the title of Goffman's (1967) book.
4. This change in power relations occurs as noted earlier by Simmel (1896) in his seminal work on superordinate–subordinate reciprocal relations, which change based on changes in number, structural changes, and position in the order of relations. Similarly, research on networks suggests that acquiring greater resources requires a large number of non-overlapping linkages such as that achieved by plugging structural holes (Burt, 1992). The manipulation of structure is also a focus of theorists in the structure–conduct–performance paradigm (Porter [1980] 1998) and game theorists (Brandenburger and Nalebuff, 1996).
5. The firm is referred to as the player in the game of innovation.
6. Rivkin (2000) applies a complex systems approach to strategy and examines the link between complexity and imitability of strategic decisions.
7. Earlier DiMaggio and Powell (1983) noted that firms often resemble each other because of imitation resulting from normative, coercive, or mimetic isomorphism; likewise, Winter and Szulanski (2001) observed that firms may use replication as strategy. Besides concurring with the view that such imitation and replications occurs, this book emphasizes that imitation is a multistage, dynamic process that takes into account the context of the environment and other lead players. It also highlights that complexity is incorporated into the system by adopting new rules that may be quite simple. In a sense, firms discover the algorithm for success in the game of international expansion and new rules are devised from observation.
8. This evolution is corroborated by the literature on growth via a 'big push' towards industrialization (Young, 1928; Rosenstein-Rodan, 1943; 1944; Murphy et al., 1989) and investment in increasing returns technologies (Romer, 1986; Arthur, 1989; Krugman, 1991). However, this literature does not focus on firm-level learning and capability development specifically.
9. Permutter (1969) cautioned however, that developing a geocentric mindset is not easy and requires commitment and resources.
10. The wholesale adoption of values and orientations associated with markets is not without problems. See discussion later in chapter.
11. The popularity of Senge's (1994) model of the learning organization was reflected in the language of Steelworks' managers and executives who used the label in speeches to justify change and transformation.
12. Audretsch (1991) distinguishes between 'entrepreneurial' and 'routinized' regimes and indicates that entrepreneurial regimes are favorable to innovative entry. In such a regime information outside the company is more important than information inside the company. Consequently, newly established firms have the innovative advantage over incumbent firms. In contrast, in routinized regimes (Nelson and Winter, 1982), the accumulated stock of non-transferable information is the product of experience within the market that the firms outside of the industry cannot possess.
13. Geographical boundaries are less relevant only when shared conceptual maps exist between boundary-crossing participants.
14. Participation in the global economy entails increasing specialization (Simmel, 1898) and division of labor both within and across firms (Durkheim [1933] 1984) and the emergence of new niches for Indian firms.
15. These included four multinational subsidiaries in the software and biotechnology industries. The data from these organizations are not reported here.
16. Foucalt (1980) uses this approach in his study of the penal policy in France and draws on instances in other societies to corroborate his views. Similarly, I draw on theories about firms in other countries to derive insights about the expansion of Indian firms.

17. Many interviewees noted the importance of a good work ethic to sustain continuous improvement and evoked the philosophy of 'karma yoga', the Indian equivalent of the 'Protestant ethic' (Weber, 1930) in describing their attitude to service and work. Some organizations also sponsored training in yoga for employees; companies had long focused on community-building to improve conditions for employees and were now able to leverage these strengths when pursuing opportunities in other developing countries.
18. Braverman (1974).
19. Foucault (1980) notes that knowledge and power are intimately related; discourse is transformed in, through, and on the basis of relations of power. The wider diffusion of methods of industrial knowledge production is thus associated with the spatial and geographical dispersion of panoptism via new forms of power and control.
20. Geertz (1978) earlier noted the difference between the 'bazaar' and 'market' economies, arguing that the bazaar economy focused on building long-term client relationships to reduce information asymmetry.

Appendix A A note on the Indian steel, construction equipment, and auto-component industries

Before economic liberalization in 1991, the Indian government exercised pervasive, continuous but variable controls on every aspect of industrial activity. A policy of import substitution in the late 1950s led to a highly protected and inward-looking regime in which the private sector played a limited role. This regime continued until the early 1980s. The focus of policy was to foster indigenous technology, reduce the role of foreign ownership, limit foreign investment, and promote manufactured exports. Exports were regarded as important to the extent that they could finance products or inputs not manufactured in India. A policy of quantitative restrictions on imports helped to protect products manufactured in India. Imports of manufactured consumer goods were banned and intermediate products and industrial spares were allowed to be imported when not available locally or for export production.

STEEL

The iron and steel industry, prized as a 'basic' industry by newly industrializing countries for providing crucial input into all forms of industrial development, is a mature, capital-intensive industry. Process technology for manufacturing basic steel is well diffused and can be readily imported on an arm's length basis in the form of engineering services and equipment (Lall, 1987). While the technology of iron and steel manufacture are embodied in capital equipment, the nature of the process necessitates complex engineering. The process of iron and steel manufacture is semi-continuous with different levels of production, raw material preparation, iron smelting, steelmaking, and the manufacture of finished or semi-finished steels. Different technologies requiring specific skills are used at each stage. Overall productivity depends on the supply of raw materials of the right quality, capacity-matching, scheduling material flows, de-bottlenecking and full utilization of installed capacities, requiring 'tacit' technical skills in process coordination, control, and productivity. Large equipment requires constant maintenance and can be significantly improved by 'minor' innovation (Lall, 1987).

The pace of technological progress in the industry is slow, with decades passing between large jumps in technology in the three major stages of iron-steel production: iron smelting, steelmaking, and processing of liquid steel. Blast furnace (BF) technology has not changed for decades although technological improvements have been made to improve the quality of the blast furnace 'burden' (material inputs) by raising the proportion of sinter, or by palletizing the ore, use of instrumentation to improve control of BF temperatures and reactions and so on. The introduction of the basic oxygen process (BOP) in the early 1950s was a major jump in steelmaking technology that represented a major advance over the existing open hearth (OH) process, which had, in turn replaced the Bessemer process. BOP has half the energy costs of OH processes, a much shorter tapping time and lower investment costs and has a minimum scale of over 1 million tonnes per year. New technology for processing liquid steel is continuous casting (CC), which combines the three stages of ingot casting, soaking pits and blooming mills, thus increasing yields, improving quality, and lowering investment and space requirements (Lall, 1987).

Indian steel production started under free market conditions; modern iron production was launched at Kulti in 1874 but the first integrated steel producer began production in 1912 with its own collieries and iron ore mining. British interests started another integrated plant, which became wholly Indian-owned by the 1960s and nationalized in 1973–74 after poor performance led to mounting losses. Indian steel was fully competitive and considerably cheaper than imported steels in the 1950s and 1960s. By the 1960s, import substitution had eliminated the threat of potential foreign competition. Local prices were restrained by price controls and imports were permitted only of special steels or of ordinary steel to make up for shortfalls. Total steel output in India rose in the 1960s when the public sector entered the industry with three integrated plants from 3.3 million to 6.23 million tonnes. By 1981–82 India was the sixteenth largest producer of steel in the world, producing 1.3 million tonnes of pig iron and 8.8 million tonnes of mild steel. There were six integrated iron and steel plants, together accounting for 92 percent of pig iron and 81 percent of mild steel production, one in the private sector, and the others grouped under a large public sector firm (other secondary iron producers and mini steel plants existed but are not discussed here; Lall, 1987).

However, the performance of the public sector integrated plants deteriorated after 1976. All public sector plants were turnkey plants gifted by Russia, the United Kingdom, and West Germany with no indigenous design content. Initial technology transfers to the public sector plants were problematic because of poor engineering, failure to take account of local raw material characteristics (poor-quality coking coal with high ash content and coke with poor caking qualities; iron ore that is difficult to beneficiate), and because the technologies transferred were sometimes not the best available at that time. Government controls kept prices down despite the government's large consumption of steel for infrastructure and industrial uses. Low prices were complemented by a system of official allocation and distribution to different users. Over time, the public sector plants ran up huge losses. In contrast, the private sector firm consistently made profits under the same system because of its own technological capabilities, but at low levels that held back large-scale investments. Nevertheless, the private sector company was unable to expand its capacity beyond 2 million tonnes from 1959–83 as the Industrial Policy Resolution of 1984 reserved all future steel capacity for the public sector; existing private producers were permitted to continue as long as their performance was 'satisfactory' with the threat of nationalization hovering in the background. Thus, modernization was held back because of the threat of nationalization and because price controls reduced profitability[1] (Lall, 1987).

The private sector firm undertook partial modernization of its plant only in 1983. In 1983, the firm employed about 60 000 people, ran its own township, and launched its first expansion after nearly 30 years. It had built up a team of dedicated professional managers and technologists along with a skilled and cohesive labor force. Equipment maintenance and project execution capabilities had been strengthened by vertical integration into equipment manufacture although this capability could only be used in-house in accordance with government licensing rules. Its maintenance shop was expanded during the mid-1960s at a cost of US$5 million into a full-blown equipment manufacturing facility with 120 designers (half were engineers) to manufacture products at about 50 percent less than the cost of imports, reducing reliance on imports during a growing foreign exchange crisis. Similarly, process improvements were made over time in all areas of operations and new steels developed through in-house R&D were regularly introduced to replace imports. Industrial engineering capabilities were also fostered in the process of continuously improving old plant and maintaining high rates of capacity utilization and product upgrading. Personnel needs were met through a comprehensively planned system that favored hiring new graduates as recruits and providing further formal and on-the-job training in-house. Promotion was mainly internal to retain skilled people and loyalty (Lall, 1987).

Liberalization of India's industrial policy led to increased modernization, expansion, and establishment of new greenfield plants using state-of-the art technologies in the 1990s by existing and new players to meet global competition (see Chapters 4 and 5 for details on how a leading firm built new capabilities). By 2007, with an economic growth rate of 9.2 percent and improvements in income, the steel industry experienced increased sales of 40.9 percent (for the first three months of 2007) in comparison with the same period in 2006 (rising from INR367465 million to INR517040 million) while operating profit for the industry grew approximately 65 percent during the same period.

Steel is one of the top products in the manufacturing sector; steel production has been reduced in the United States and Europe, to match slowing demand. Nevertheless, global steel production is up 10.4 percent. In fiscal year 2006, China was the largest producer of steel (421.5 million tonnes or 34 percent of global production) followed by Western Europe (199.5 million tonnes, 16 percent) and North America (132.6 million tonnes or 10.7 percent).

Total crude steelmaking capacity is over 34 million tonnes in 2007. India is the eighth largest producer of steel in the world and produces a variety of grades meeting international quality standards. Global markets have accepted Indian steel because Indian HR (hot rolled) products are classified by 'World Steel Dynamics' as tier-2-quality products along with South Korea and the United States, while the EU and Japanese products are categorized as tier-1.

Demand for steel is likely to continue to rise because of the rising demand for automobiles, increased housing and construction activities and boom in infrastructure. Of the steel produced, 60 percent is consumed by the construction sector, 28 percent by the automobile sector, and the rest (12 percent) by other industries. Investment in the steel industry accounts for nearly 11 percent of India's gross domestic product (GDP) and nearly 50 percent of its gross capital formation. World steel industry trends show an upward trend largely fuelled by demand in China. Although the Chinese steel industry is expected to grow, China is not likely to be the lowest-cost producer; hence prices are expected to rise in 2007. Domestic production in India is expected to rise by 8 percent to 51.8 million tonnes in 2007–08 (CMIE, 2007; Cygnus Business Consulting and Research, 2007).

CONSTRUCTION EQUIPMENT

Construction and mining equipment cover a variety of machinery such as hydraulic excavators, wheel-loaders, backhoe loaders, bulldozers, dump trucks, tippers, graders, pavers, vibratory compactors, cranes, forklifts, drills, scrapers, motor graders, rope shovels, and so on. They perform functions for mining and construction activity such as preparation of the ground, excavation, material haulage, dumping, material handling, and road construction. India has a few medium-sized and large companies in the organized sector in this industry, as technology barriers restrict the entry of small and medium enterprises. Before the 1960s, domestic requirements of mining and construction equipment were met through imports. Domestic production began in 1964 with the establishment of a public sector enterprise in South India to manufacture dozers, dumpers, graders, scrapers, and so on for defense requirements with technical collaborations with companies in the United States and Japan. From 1969 onwards, a few private sector companies also began manufacturing hydraulic excavators with technical assistance from Japan, the United Kingdom and the United States. Multinational companies are the most recent entrants; however, most of the technology leaders are present in India as joint venture companies or have established their own manufacturing facilities or marketing companies.

Public limited companies, including public sector units, comprise 71 percent of this sector and 29 percent are private limited firms. Most (75 percent) are involved in the activities ranging from design and engineering, manufacturing, servicing, and commissioning. A few companies act as selling agents for multinational players while others manufacture and import complete equipment from companies abroad. The international trend in the earthmoving and mining segment is one of consolidation; in addition, some international companies are looking at using Indian operations to meet demand in South and Southeast Asia. The total market size of three product groups (earthmoving machinery, construction machinery, and cranes) amounts to INR42 320 million (Government of India, c. 2006). Demand for construction equipment is linked economic growth (see Figure A1) and investment in infrastructure (for example, projects such as the Golden Quadrilateral highway project in 2002–03); the sector has experienced double-digit growth in sales turnover for 2003–05 and 33 percent growth in 2004. Domestic demand in 2004–05 was INR63 billion and demand for 2005–06 was estimated at greater than INR70 billion. Exports amounted to approximately INR2.8 billion in 2003–04 and INR3.3 billion in 2004–05 (ibid).

Despite access to the latest technology through joint ventures and the presence of manufacturing capabilities in India, the equipment manufactured domestically does not use these technologies partly because of low volumes, uncertain demand, and cost of the latest technology. Also as the field study suggests, R&D spending by domestic firms is much lower than spending by multinational firms (see Figure A2).

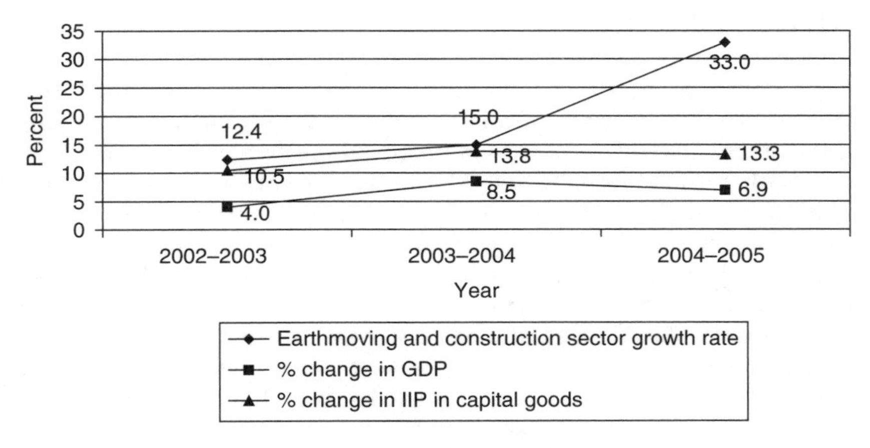

Source: Government of India, Department of Heavy Industry.

Figure A1 *Growth of the construction equipment industry correlated with economic growth*

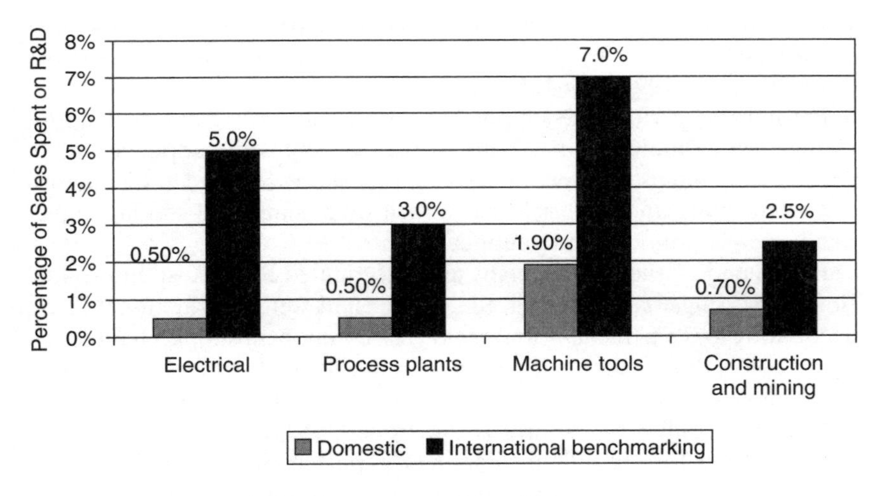

Source: Government of India, Department of Heavy Industry.

Figure A2 *R&D (c. 2004–05) spending by domestic and international companies in construction equipment and mining and the capital goods sector*

AUTO-COMPONENTS

The automobile industry in India underwent a transformation in the 1990s. At the beginning of the decade, production levels were modest (India produced about 209 000 cars) and multinational involvement was limited. A link with Suzuki, forming the Maruti-Suzuki company (now Maruti Udyog) helped in this transformation by capturing 70 percent of passenger car sales by the early 1990s, dramatically reducing the market share of the dominant Hindustan Motors whose 'Ambassador' model had been the largest selling model for decades.

The Suzuki-Maruti plant, located outside Delhi, developed a network of suppliers in the early 1990s. Some of these were joint ventures and others were independent domestic firms. Suzuki worked with both types of suppliers to establish international best practice and achieve quality and productivity.

From the 1990s onwards, a wave of multinational firms entered India; they were required to achieve a high level of domestic content (usually 70 percent within three years). Achieving this target required switching from reliance on imported components to sourcing from local vendors, giving the automobile manufacturers a strong incentive to work with tier-1 suppliers to meet quality standards at an acceptable price (Sutton, 2004). Car production increased by the end of the decade by a factor of 2.5 (from 209 000 units in 1991 to 564 000 in 2001).[2]

During this period, the supply chain had also undergone tremendous change. While multinational firms worked closely with suppliers, Indian firms, facing intense competition for market share, responded by upgrading productivity and quality levels in their own plants and seeking higher quality levels from their own suppliers.

By the end of the decade, eight firms accounted for almost all production of passenger cars in India. Six of the eight were multinational firms, accounting for 85 percent of units sold. The component supply chain developed rapidly during the same period with the value of component production almost doubling from 1997–2001 (from US$2406 million to US$4203 million) and value of exports rising from US$299 million to US$555 million. Of the top ten Indian component exporters, six were multinational joint ventures while three formed part of an Indian group. A well-developed supply chain is characterized by a couple of key components (such as the cylinder head and cylinder block) being manufactured in-house while a central group of key components, assemblies or sub-assemblies (for example, the engine mounting, crankshaft, and transmission) may be outsourced or made in-house. Finally, a group of less central components (such as pistons, exhaust system, and bumpers) are normally outsourced (ibid.).

Table A1 Production of passenger cars in top 12 countries in 2004 (in units)

Rank	Country	Production	Growth Rate (%)
1	Japan	8 720 385	3
2	Germany	5 192 101	1
3	USA	4 229 625	−6
4	France	3 227 416	0
5	Korea	3 122 600	13
6	Spain	2 399 374	0
7	China	2 316 262	15
8	Brazil	1 756 166	17
9	UK	1 646 881	−1
10	Canada	1 335 464	0
11	India	1 178 354	30
12	Russia	1 109 958	10
	World Total	44 099 632	5.1

Source: ACMA (Automobile Component Manufacturers Association of India), New Delhi, India.

The supply chain in the Indian auto-component industry developed rapidly at the level of car makers and tier-1 suppliers (ibid.). However, best practice techniques are permeating down to tier-2 suppliers more slowly and unevenly. While development was driven by the presence of international car manufacturers, component exports are driven equally by multinational and domestic firms. Half of the top ten Indian firms are domestic firms and three of these belong to a single domestic industrial group. Supply chain development in the auto-component industry allows domestic car makers to outsource more effectively and achieve cost reductions while maintaining quality levels (see Tables A1 and A2 for details on world car production and quality certification in auto-components). Leading component producers also use highly capital-intensive techniques in these low-wage environments largely to achieve high levels of quality control in the production process. Additionally, Sutton (2004) notes that deployment of highly qualified individuals and extensive provision of in-house training for shop-floor operations also helps to achieve high quality standards necessary for export success.

The auto-component industry grew by 15 percent in 2006 to US$10 billion and exports rose by 29 percent to US$1.8 billion (SSKI, 2006; see Figure A3). The Indian auto-component industry manufactures the entire range of parts required by the domestic automobile industry and currently employs about 250 000 people (Government of India, Department of Heavy Industry, 2006, p. 18, 26).

Table A2 Quality certification and best practice in the auto-component industry

Number of Companies	Certification	Types of Modern Shop-floor Practices Suppliers Adopted
456	ISO 9000	1 5-S, 7-W
248	TS 16949	2 Kaizen
136	QS 9000	3 TQM
129	ISO 14001	4 TPM
32	OHSAS 18001	5 Six sigma
9	Deming prize winners	6 Lean manufacturing
4	JIPM award winners	
1	Japan Quality Medal winner	

Note: Total ACMA member companies: 498.

Source: ACMA, (Automobile Component Manufacturers Association of India), New Delhi, India.

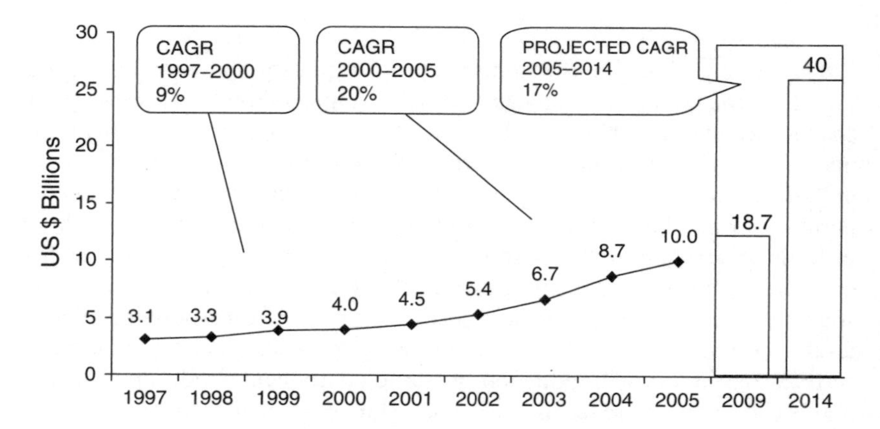

Note: CAGR = Compound Annual Growth Rate.

Source: ACMA- McKinsey Report.

Figure A3 Growth of auto-component industry

Exports of leading auto-component companies were affected by the slower than expected pace of outsourcing by global vendors and margins were under pressure from higher input costs; nevertheless, the Indian auto-component industry is expected to achieve US$40 billion in size by 2015 with the share of exports estimated at 50 percent (SSKI, 2006). Key export markets include the United States and Europe together constituting 62 percent of total auto-component exports from India. Faced with competition domestically, leading auto-component vendors from the United States and Europe are searching for suitable companies in low-cost locations like India, China, and Thailand for outsourcing component requirements. While Indian companies have achieved high quality, the scale of operations is small. Of the top 120 auto-component companies, 75 percent have revenues less than US$50 million, while revenues of only 15 percent of the companies exceed US$100 million. Indian companies are currently awarded global orders of less than US$50 million (ibid.).

NOTES

1. However, one major private sector firm was allowed to pursue metallurgical consultancy to compete with one in the public sector. Heavy steel plant was reserved for the public sector but ancillary and rolling mill equipment was left open to private firms.
2. During the financial year 2005–06, the Indian automobile industry produced more than 9.7 million vehicles amounting to almost USD 28 billion. India produces about 1 300 000 passenger vehicles, 400 000 commercial vehicles, 7 600 000 two-wheelers and about 300 000 tractors per annum. India is the second largest market for two-wheelers in the world. However, in value terms, the market for passenger cars and commercial vehicles is greater than that for two-wheelers (Government of India, 2006, p. v, 7).

Appendix B Indian software industry: historical background

The diffusion of information technology (IT) into all other industrial and service sectors makes it one of the most critical technologies affecting economic growth in developing countries (World Bank, 1992). Failure to introduce new information technologies is likely to result in inefficient administrative and production methods. While IT includes both hardware and software technologies, software is vital because other technologies cannot function without it. Software is an important component of overall value within information technologies and is becoming a pervasive technology embodied in a vast and diversified range of products and services (Gaio, 1989).

The development of a local software industry is seen as a necessity for developing countries to be able to adapt software technology to suit their particular local needs. Software production is also seen as the best entry point for developing countries into the IT production complex because of lower entry barriers and capital intensity, greater labor intensity and a lower rate of obsolescence (for some types of software) and fewer economies of scale. Labor intensity of production offers an opportunity to developing countries compared with other production processes. Thus, developing countries' production and use of software is becoming more intense and India, China, Brazil, Mexico, Singapore, Hong Kong, Taiwan, South Korea, and the Philippines all have software industries of note with annual growth rates of 30 to 40 percent not being uncommon (Heeks, 1996).

Other interesting features of software include the fact that it is intangible, modifiable after initial production to create a new product, and with a lack of any clear distinction between production tools and final product. The production process is highly skills-intensive, while certain types of production rely on labor mobility and a rapidly growing world market.

India has one of the longest-established and largest software industries in developing countries. It has had a software policy since 1970, longer than other developing countries. India's software policy has continually stressed exports but was liberalized in the mid-1980s when software was identified as a thrust area (*Electronics Information and Planning*, 1985).

The Indian experience also differs because of the industry's export focus. Contrary to the view that a strong domestic-oriented industry is necessary before moving into exports (Fialkowski, 1990; Schware, 1992 in Heeks, 1996; Porter, 1990), the growth of the Indian software industry seems to have been driven primarily through exports to foreign markets.

SOFTWARE, HARDWARE, AND SOFTWARE PRODUCTION

Technologies involved in computing are normally divided into hardware and software. Hardware is the mechanical, magnetic, electronic, and electrical devices that make up a computer, while software is the instructions, programs, or suite of programs that are used to direct the operations of a computer or other hardware, plus associated documentation (Meadows et al., 1987). Software consists of algorithms or 'recipes' for processes that can be executed via programs. Programs can be subdivided into subroutines or modules, each of which consists of a series of instructions.

Software can be produced merely by writing the instructions with pen and paper or a computer; however, testing and using the software require a computer. Hardware as well as labor are necessary inputs to the software production process. Instructions that make up software are written in a variety of computer languages such as C, Visual Basic, C Sharpe, JAVA. Using a programming language, software is initially written in a human-readable form, called source code. This is then translated into a computer-readable string of ones and zeros, called object code, which operates the computer. In addition, software tools exist that can generate software instructions automatically. Thus, software tools and languages are also inputs to the software production process.

The software production process is usually managed by breaking up the process into a series of steps: analysis of the problem, design of the software, coding (writing) the software, testing the software, delivery, and subsequent maintenance.

Software produced has been divided into two basic categories: applications software is 'programs . . . designed to carry out specific tasks or applications as distinct from systems software which controls the operation of the total computer system' (ibid.).

Software may be produced in a standardized form for general sale to a large number of users, in which case the product is known as a software package; or it may be produced for a single specific customer, in which case the product is known as customized software. Customized software is also used to describe standard packages modified to suit a particular client's needs. While software packages are viewed as goods and customized software as a service, in practice these distinctions are often blurred.

THE SOFTWARE INDUSTRY

The companies producing software are collectively referred to as the software industry. This term encompasses companies or company divisions that earn the majority of their revenue from software consultancy services, software packages, and outsourcing of business processes.

Science and technology have always had an important role within India's industrial development with the importance of technology being stressed since the earliest post-independence days to ensure self-reliance and building up of local technological capability. In the 1950s and early 1960s, technology import policy was quite liberal to provide a base on which to build local capabilities but there was greater selectivity from the mid-1960s onwards partly in response to China's nuclear test and the US export embargo in 1965. In mid-1970 a Department of Electronics was established to oversee electronics and computer-related industries including the software industry and software industrial policy. Policies were partly liberalized and it was made clear that software was eligible for export incentives such as the location of production in export processing zones (EPZs; Heeks, 1996, p. 43). Some restrictions were introduced again between 1980 and 1984 to curb imports of computers primarily for domestic use, resulting in delays and confusion (*Economic and Political Weekly*, 1984).

Science and technology were pushed to the fore under Rajiv Gandhi in the mid- and late 1980s and there was a move away from import substitution towards modernization through import liberalization and export orientation (Lall, 1987). Although policy changes mainly affected hardware, some were directed towards software such as lowering import duties on hardware and software. Software was recognized as an industry and was delicensed and large companies (those covered by the Monopolistic and Restrictive Trade Practice [MRTP] Act) and those with up to 40 percent foreign equity holdings (covered by Foreign Exchange Regulation Act) were allowed to become software producers. By 1986 the import of hardware was simplified and import of software delicensed (changed from quota to tariff protection) so that anyone could import it if they paid the 60 percent import duty. The Department of Electronics' software development agency was established in 1986 to formulate and implement software-related policy measures and promote the software industry. An insurance scheme introduced in 1987 to cover clients of Indian software companies against malpractice and export shipment credit and credit guarantees were introduced. Venture capital funding for software companies also became available as did overseas telecommunication links. The idea of software technology parks (STPs) was also introduced. The Electronics and Computer Software Export Promotion Council was created in 1988 to increase exports of electronics goods and software through marketing.

In the 1990s these changes were further accelerated and new policies in 1993 were much more oriented towards globalization, allowing freer technology imports, reductions in duties for software imports, and more automatic approval of technology transfer (Heeks, 1996).

The government's principle objective was to earn foreign exchange through software exports to compensate for the costs of importing hardware and electronics items (Heeks, 1996). Software was considered suitable for export promotion because of the large growing world market, the perceived low investment requirements and the availability of skilled, English-speaking low-paid workers (Heeks, 1996). It was an export thrust area with a target of 60 percent value addition compared with 20–33 percent for most other exporting industries. Thus, production for the domestic market was neglected. The only policies directly targeting the domestic software market were those pertaining to the encouragement of R&D and the floating of local tenders. Thus, compared with other industries, the software industry was virtually the only one that was primarily export-oriented both in practice and policy (Sridharan, 1989, p. 53).

The evolution of the software industry in India is linked with the development of the Indian hardware industry. Until the mid-1960s hardware and software were provided by multinational companies like IBM and ICL. The software they sold had been developed overseas. As in the West, since computer manufacturers could not provide the full range of application software (Kaplinsky, 1987), software development became the purview of in-house developers producing software for their own organizations. As the number of commercial organizations using computers grew, software development began to be contracted to other organizations such as management consultancies. Thus, a domestic market for software emerged.

A number of companies began to import hardware on the condition that they export software beginning in 1974 and the data-processing departments of some large companies and software groups of some Indian hardware manufacturers began trying to sell their in-house software. As the revenue-generating potential of software became apparent, firms began to make their software units more outward-looking, sometimes creating a separate company. The departure of IBM in 1978 provided an added boost, with several ex-IBM employees setting up small software companies, and led to the growth of the software industry.

Exports began to grow after 1981 and by the mid-1980s, multinational companies began to consider India as a software development source and as a market for software products. From the late 1980s, multinational and local interest in exports increased and a number of large Indian firms created software divisions. By the mid-1990s, hardware manufacturers had also begun to focus on software exports.

Appendix C Evolution of biotechnology in India

While the biotechnology industry in India is of relatively recent origin, it has its roots in government efforts to promote research in biotechnology beginning with the establishment of the Department of Biotechnology (under the Ministry of Science and Technology) in 1986 to coordinate talents, materials, resources, and budgetary provisions. Thus, biotechnology has historically been a government-sponsored effort with little private participation in investment although recent trends show that private sector participation is increasing (Ghose and Bisaria, 2000).

India confronts problems raised by a fast-growing population, degradation of the environment, destruction of forest cover, inadequate health care and nutrition, and damage of agricultural land. Many of these problems can be addressed by the application of available knowledge in frontier technologies such as biotechnology. Thus, biotechnology is of great interest to developing countries like India because of the potential for stimulating agricultural productivity by increasing crop yields and reducing biotic and abiotic stresses. Similarly, application of biotechnology has the potential to change the production profile of the industrial sector. As the growth of knowledge-based sectors is dependent on strong institutional support for science and technology and R&D, the historical development of institutions pertaining to biotechnology is briefly outlined (Chaturvedi, 2002).

EDUCATION AND TRAINING

Before independence in 1947, scientists and academics were engaged in intellectual advancement primarily for self-satisfaction and were funded by the government with no industry involvement. While need-based research was not pursued, there were some eminent world-class thinkers in universities such as the scientist J.C. Bose, a radio-physicist who later shifted to botany and quantified the plant's ability to respond to electrical signals.

Post-independence, the need to formulate an appropriate national policy to build up biotechnology and the requisite personnel were recognized and funds were budgeted to initiate research. A National Biotechnology Board was constituted in 1982 under the auspices of the Ministry of Science and Technology. Other events such as the establishment by the United Nations Industrial Development Organization (UNIDO) of one of the centers of its International Center for Genetic Engineering and Biotechnology in New Delhi, the hosting of the International Biotechnology Symposium at New Delhi in 1984, and the International Genetics Congress contributed to laying the foundation of the biotechnology initiative in India. The National Biotechnology Board was soon converted into a new Department of Biotechnology.

Education and training had started as early as 1964 with Bachelor's programs in food technology, biochemical engineering, and fermentation to cater to the needs of the processed food industry in Calcutta, and subsequently at Kanpur and Mumbai. The inadequacy of these programs to meet growing needs led to the introduction of academic training and research programs at the Indian Institute of Technology (IIT) Delhi in 1969. Substantial scientific and technical support was obtained through an Indo-Swiss collaboration between IIT Delhi and the Swiss Federal Institute of Technology, Zurich, which began in 1974 and was phased out in 1985 (Ghose and Bisaria, 2000). It finally evolved into a world class-center of Biochemical Engineering Research and led to the establishment of the first academic Department of Biochemical Engineering and Biotechnology in 1993. Since then other IITs have also established such departments.

The Indo-Swiss collaboration was extended to four new Indian scientific institutions in 1988 and two other partners were integrated in 1995. The collaboration focused on the development of sustainable scientific and technological capabilities of R&D institutions for product development and technology transfer. Projects were selected based on the criteria of scientific quality, significance and feasibility, collaborative research between institutes of both countries, feasibility of technology transfer, commercialization possibilities, legal and ethical aspects, and compliance with the guidelines of the Swiss and Indian regulatory bodies. The projects were considered Indian with largely Swiss support and ranged from human health, animal husbandry, microbial processes, products for agriculture and the pharmaceutical industry. The cost of the program was shared between collaborators according to the bilateral agreement.

After the Department of Biotechnology was created, other universities and scientific institutions were given financial assistance to create essential facilities to conduct biotechnology training programs at the Master's and PhD levels, to provide academic training of faculty at many universities abroad as well as training of technicians in selected laboratories in the country. Currently, almost all universities, IITs, and the Indian Institute of Science offer training in biotechnology. Other governmental agencies have in-house training in their respective disciplines. In addition there are some autonomous research institutes that provide training in specialized sectors of biotechnology. Also, almost all universities offer courses in life sciences, biochemistry, biophysics, molecular biology, genetics, microbiology, zoology, botany, and chemical engineering, leading to degrees in respective disciplines.

Trained personnel are chiefly engaged in R&D, production, and quality control. In medical, agricultural, and allied establishments the number of trained R&D scientists far exceeded production personnel, similar to what was generally observed in the United States, Europe, and Japan (Srivastava, 1995). However, contributions from trained personnel to industrial biotechnology were insufficient because approximately 50 percent of qualified people migrate to the United States and Europe (Ghose, 1998), industry's hesitation to absorb indigenous know-how, reluctance of MNCs to participate in India, and confusion of how to handle intellectual property of biotechnology products.

In addition, a national network of biotechnology information exchange and retrieval was established by the Department of Biotechnology in 1989. The apex center coordinates global network activities and provides bioinformatics and biocomputing services to researchers engaged in biology and biotechnology R&D and manufacturing activities across the country. The department also supports a number of repositories for conservation of living organisms for various sectors of biotechnology such as agriculture, health care, animal husbandry, and industry. International collaborations were also established with several countries in areas other than education and training. During 1987–98, more than 20 collaborative agreements in biotechnology were signed between India and countries like Switzerland, the United States, China, France, Germany, the United Kingdom, Sweden, G-15 countries, and Russia. For example, Indo-French scientific collaboration was pursued through a center established to promote research through joint seminars, workshops, and symposia of current interest.

AREAS OF BIOTECHNOLOGY RESEARCH

The most important areas of biotechnology research include those devoted to agriculture and health care. Agricultural biotechnology includes plant tissue culture to improve crop varieties and yields of priority crops such as rice, wheat, mustard, chickpeas, mungbeans, peas, and cotton. Noteworthy achievements include the development of triploid plants through endosperm culture, a novel technique of test tube fertilization to overcome incompatibility in plants exhibited in wild crossing and flowering of bamboo. In addition, progress has been made in the biocontrol of plant pests through the development of cost-effective, commercially viable technologies for biocontrol agents like baculoviruses, parasitoids, antagonistic fungi, and bacteria for use in managing pests and diseases of economically important crops. Tissue culture techniques applied to tree and woody species have also been developed at various institutions to help in the mass production of disease-resistant species. Likewise, medicinal and aromatic plants are being micropropagated to conserve their germplasm and harness their economic potential. For example, *Taxus sp.*, a source of the anti-cancer drug Taxol has been studied at various centers. Finally, to take advantage of India's rich biodiversity in two spots in the Northeast Himalayas and South Western Ghats, a major initiative on bioprospecting involving 13 collaborating institutions in India was launched in 1997.

In medical biotechnology, major research areas include DNA transactions, protein structure and function, protein engineering, reproductive endocrinology, developmental biology, and ecology. Other important areas include reproductive biology, work on food and environmental allergens, and immunological studies on a variety of infectious diseases and DNA vaccines. Research activities can thus be grouped under infectious diseases, drug and molecular design, genetic disorders, gene targeting, and genetic diversity (Ghose and Bisaria, 2000). Examples of applied projects include male fertility regulation, anti-tubular drug screening and a new drug target for malaria. Some of these projects have been developed in collaboration with industry such as a peptide diagnostic kit for HIV, DNA-based test procedures for genetic disorders, and a recombinant hepatitis B vaccine.

Research efforts are also directed towards food biotechnology (applicable to food processing), animal biotechnology, and seribiotechnology (biotechnology for the production of silk, which continues to remain a sought-after commodity from India), and environmental biotechnology, which is applicable to environmental pollution control and treatment of domestic wastes and industrial effluent. Finally, in the area of industrial biotechnology, which pertains to the production of products like alcohol, biopesticides, drugs, antibiotics, enzymes, vaccines, some organic molecules, and bioreactors, achievements include production of alcohol using ethanol-tolerant strains of yeast, production of biogas, and production of glucose and dextrose by enzymatic liquefaction of starch, mainly tapioca (cassava). R&D spending of pharmaceutical companies was about 1.3 percent of turnover (just over INR1200 million compared with the total turnover of INR90 billion). See Figure C1 for the composition of the Indian biotechnology sector in 2005–06.

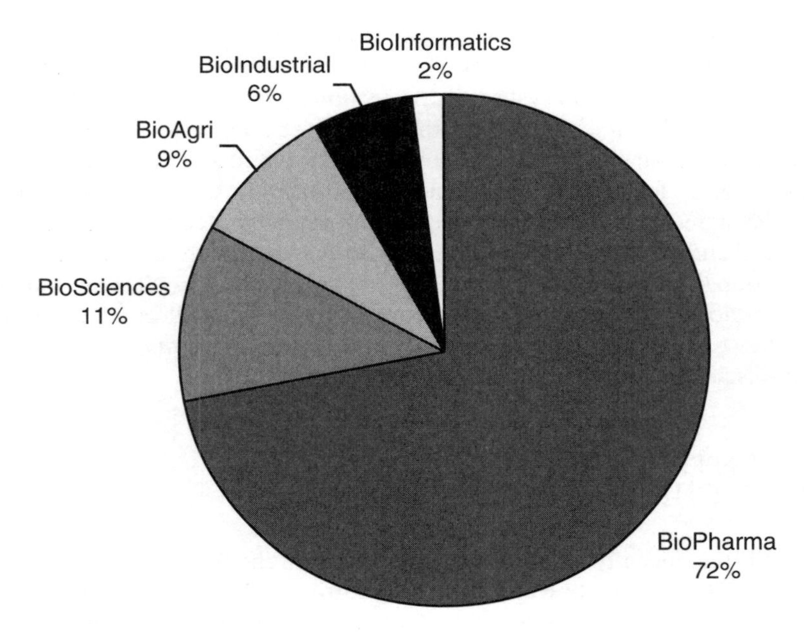

Source: Confederation of Indian Industry.

Figure C1 Composition of the Indian biotechnology sector (2005–06)

INTELLECTUAL PROPERTY RIGHTS

The intellectual property regime also played a role in the development of the biotechnology industry. India signed the Final Act of the Uruguay Round of multilateral trade negotiations in 1994 in Morocco. This act of the General Agreement on Tariffs and Trade (GATT) covered inter-alia issues related to intellectual property rights. The seven areas of intellectual property rights covered by Trade Related Intellectual Property Rights (TRIPS) are trademarks, trade secrets, industrial designs, copyrights, integrated circuits, geographical indication, and patents. While Indian laws, regulations, administrative procedures, and the judicial system are equivalent to the rest of the world on the first six, in the area of patents, Indian laws differ substantially from the provisions of the WTO. Unlike the WTO, which provides product patents in all branches of technology, Indian laws did not permit product patents in drugs, food, and chemicals; instead, process patents were permitted. While the WTO provides patents on microorganisms and microbiological processes, Indian patent laws do not permit patenting any life form though patents based on microbial processes are permitted. In addition, while the WTO requires protection of plant varieties by patents or an effective 'sui generis' system or by any combination thereof, there is no system for protection of plant varieties in India.

Finally, while all patents under the WTO have a uniform duration of 20 years, the Indian system provides seven years for food and pharmaceuticals and 14 years for others (Ghose and Bisaria, 2000). As a signatory to the TRIPS agreement, India must implement patent protection on pharmaceutical and biotechnology products by 2005. Hence, steps have been taken to rationalize policies to conform to WTO provisions. Several Indian biotechnology companies have managed to cross intellectual property hurdles to work with international partners through confidentiality and non-disclosure agreements (CII, 2003).

INCENTIVES TO STIMULATE THE BIOTECHNOLOGY SECTOR

In addition, incentives to increase investments in the Indian biotechnology industry and stimulate R&D have been introduced in recent years. Some examples are 100 percent foreign equity in manufacturing all drugs except recombinant DNA products and cell-targeted therapies, import duty exemptions on certain goods, weighted tax deductions of 150 percent on R&D expenditures, three-year excise duty waiver on patented products, exemption from customs and excise duties on all drugs and materials imported or produced domestically for clinical trials, and removal of minimum export obligations.

The Foreign Trade Policy 2004–09 permits the establishment of biotechnology parks across the country along the lines of Software Technology Parks and envisages that all facilities available will be 100 percent export-oriented units. Such incentives are expected to bring biotechnology on par with IT. Moreover, state governments in India have recognized the potential of biotechnology to fuel economic development and have taken initiatives to develop bioclusters based on academic and entrepreneurial strengths. These bioclusters include biotech parks, biotech policies, fiscal incentives and centers of biotech excellence, biotech development funds, and incubators. Among the fastest-growing clusters are those in Andhra Pradesh (centered around Hyderabad), Karnataka, (centered around Bangalore), and Maharashtra (centered around Pune) while others are emerging in Tamil Nadu and Orissa (CII, 2003). Opportunities are envisaged in all areas of biotechnology.

References

Acs, Zoltan J., David B. Audretsch and Maryann P. Feldman (1994), 'R & D spillovers and recipient firm size', *Review of Economics and Statistics*, **76** (2), 336–40.

Aharoni, Yair (1966), *The Foreign Investment Decision Process*, Division of Research, Graduate School of Business Administration, Boston, MA: Harvard University.

Almeida, Paul and Bruce Kogut (1999), 'Localization of knowledge and the mobility of engineers in regional networks', *Management Science*, **45** (7), July, 905–17.

Aoki, M. (1990), 'Toward an economic model of the Japanese firm', *Journal of Economic Literature*, **28** (1), 1–27.

Arora, Ashish (1996), 'Contracting for tacit knowledge: the provision of technical services in technology licensing contracts', *Journal of Development Economics*, **50** (2), 233–56.

Arora, Ashish and Jai Asundi (1999), 'Quality Certification and the Economics of Contract Software Development: A Study of the Indian Software Industry', Working Paper 7260, Cambridge, MA: National Bureau of Economic Research.

Arora, Ashish, Andrea Fosfuri and Alfonso Gambardella (2001), 'Markets for technology and their implications for corporate strategy', *Industrial and Corporate Change*, **10** (2), 419–51.

Arrow, K. (1974), *The Limits of Organization*, New York: Norton.

Arthur, W. Brian (1989), 'Competing technologies, increasing returns, and lock-in by historical events', *The Economic Journal*, **99** (394), 116–31.

Association for Equipment Manufacturers (2004–05), *Outlook for Construction Equipment Business*. Available at: http://www.aem.org/Trends/AnnualCEB:z/PDF/Outlook_ 2004–5.pdf. Accessed 11 October 2007.

Audretsch, David B. (1991), 'New-firm survival and the technological regime', *Review of Economics and Statistics*, **73** (3), 441–50.

Augier, Mie Elizabeth and Saras Sarasvathy (2003), 'Management as a Science of the Artificial', Paper presented at the Academy of Management Conference, Seattle.

Baldwin, Carliss Y. and Kim B. Clark (2000), *Design Rules*, Cambridge, MA: MIT Press.

Barney, J.B. (1991), 'Firm resources and sustained competitive advantage', *Journal of Management*, **17** (1), 99–120.

Bartlett, Christopher A. and Sumantra Ghoshal (1988), 'Organizing for worldwide effectiveness: The transnational solution', *California Management Review*, **31** (1), 54–74.

Bleeke, Joel and David Ernst (1991), 'The way to win in cross-border alliances', *Harvard Business Review*, **69** (6), 127–35.

Bowman, E.H. and D. Hurry (1993), 'Strategy through the option lens: an integrated view of resource investments and the incremental-choice process', *Academy of Management Review*, **18** (4), 760–82.

Brandenburger, Adam M. and Barry J. Nalebuff (1996), *Co-opetition*, New York: Doubleday/Currency.

Braverman, H. (1974), *Labor and Monopoly Capital: The Degradation of Work in the Twentieth Century*, New York: Monthly Review Press.

Brown, John Seely and Paul Duguid (1991), 'Organizational learning and communities of practice: toward a unified view of working, learning and innovation', *Organization Science*, **2** (1), February, 40–57.

Brown, John Seely, Allan Collins and Paul Duguid (1989), 'Situated cognition and the culture of learning', *Educational Researcher*, **18** (1), January–February, 32–42.

Buckley, P. and M.C. Casson (1976), *The Theory of the Multinational Corporation*, London: Macmillan.

Burt, Ronald S. (1992), *Structural Holes: The Social Structure of Competition*, Cambridge, MA and London: Harvard University Press.

Campbell, D.T. and Julian C. Stanley (1966), *Experimental and Quasi-experimental Designs for Research*, Chicago, IL: Rand McNally.

Cantwell, J. (1989), *Technological Innovation and Multinational Corporations*, New York: Basil Blackwell.

Casson, Mark (1997), *Information and Organization: A New Perspective on the Theory of the Firm*, New York: Oxford University Press.

Chandler, Alfred D. (1962), *Strategy and Structure*, Cambridge, MA: MIT Press.

Chandler, Alfred D. (with the assistance of Takashi Hikino) (1990), *Scale and Scope: the Dynamics of Industrial Capitalism*, Cambridge, MA: Belknap Press of Harvard University.

Chang, Sea Jin (2003), *Financial Crisis and the Transformation of Korean Business Groups*, Cambridge, UK: Cambridge University Press.

Chaturvedi, Sachin (2002), 'Status and Development of Biotechnology in India', RIS Discussion Paper No. 28/2002: Research and Information System for the Non-aligned and Other Developing Countries.

Chesbrough, Henry (2003), *Open Innovation*, Boston, MA: Harvard Business School Press.

Clark, K.B. and T. Fujimoto (1991), *Product Development Performance*, Boston, MA: Harvard Business School Press.

CMIE (2007), *Monthly Review of the Indian Economy, June*, Mumbai, India: Centre for Monitoring the Indian Economy, p. 35.

Cohen, Wesley M. and Daniel A. Levinthal (1990), 'Absorptive capacity: a new perspective on learning and innovation', *Administrative Science Quarterly*, **35** (1), Special Issue: Technology, Organizations, and Innovation (March), 128–52.

Cohendet, Patrick and W. Edward Steinmueller (2000), 'The codification of knowledge: a conceptual and empirical exploration', *Industrial and Corporate Change*, June, **9** (2), 195.

Coleman, James S. (1988), 'Social capital in the creation of human capital', *American Journal of Sociology*, **94**, Supplement S95–S120.

Confederation of Indian Industry (CII) (2003), *Biotechnology*, New Delhi, India: Confederation of Indian Industry.

Coomaraswamy, Ananda K. (1965), *History of Indian and Indonesian Art*, New York: Dover Publications.

Cowan, R.P., P.A. David and D. Foray (2000), 'The explicit economics of knowledge codification and tacitness', *Industrial and Corporate Change*, **9** (2), 211–53.

Cygnus Business Consulting and Research (2007), *Quarterly Performance Analysis of Companies (January–March, 2007): Indian Steel Industry*, Hyderabad, India: Cygnus Business Consulting and Research.

Daft, R.L., R.H. Lengel and Linda K. Trevino (1987), 'Message equivocality, media selection and manager performance: implications for information systems', *MIS Quarterly*, **11** (3), September, 355–66.

Dahlman, C.J. and Larry E. Westphal (1982), 'The Meaning of Technological Mastery in Relation to the Transfer of Technology', World Bank Reprint Series, No. 217, Washington, DC: World Bank.

Darnay A.J. (ed.) (1993), *Manufacturing USA*, Detroit, MI: Gale Research.

Davidson, William (1983), 'Structure and performance in international technology transfer', *Journal of Management Studies*, **20** (4), 453–65.

Davidson, William and D.G. McFetridge (1985), 'Key characteristics in the choice of international technology transfer mode', *Journal of International Business Studies*, **16** (2), Summer, 5–21.

Davies, Howard (1977), 'Technology transfer through commercial transactions and the theory of the firm', *Journal of Industrial Economics*, **26** (2), December, 161–75.

Davis, Jason P., Kathleen M. Eisenhardt and Christopher B. Bingham (2007), 'Developing theory through simulation', *Academy of Management Review*, **32** (2), 480–99.

Dewey, J. (1988), *The Middle Works of John Dewey*, Carbondale, IL: SIU Press.

Dierickx, I. and K. Cool (1989), 'Asset stock accumulation and sustainability of competitive advantage', *Management Science*, **35** (12), 1504–13.

DiMaggio, Paul J. and Walter W. Powell (1983), 'The iron cage revisited: institutional isomorphism and collective rationality in organizational fields', *American Sociological Review*, **48** (2),147–60.

Dixit, Avinash and R.S. Pindyck (1994), *Investment under Uncertainty*, Princeton, NJ: Princeton University Press.

Dore, Ronald P. (1973), *British Factory, Japanese Factory: The Origins of National Diversity in Industrial Relations*, Berkeley, CA: University of California Press.

Dore, Ronald P. and Mari Sako (1989), *How the Japanese Learn to Work*, New York: Routledge.

Dunning, John H. (1981), 'Alternative Channels and Modes of International Resource Transmission', in T. Sagafi-nejad et al. (eds), *Controlling International Technology Transfer: Issues, Perspectives, and Policy Implications*, New York: Pergamon, Ch. 1.

Durkheim, Emile ([1933]1984), *The Division of Labor in Society*, New York: Free Press.

Economic and Political Weekly (1984), 'Technological goals forgotten', *Economy and Political Weekly*, **1** (December), 2018–19.

Eisenhardt, K.M. (1989), 'Building theories from case study research', *Academy of Management Review*, **14** (4), October, 532–50.

Electronics Information and Planning (1985), 'Report of the study team on electronics exports', *Electronics Information and Planning*, **12** (9), 529–37.

Enos, J.L. and W.H. Park (1988), *The Adoption and Diffusion of Imported Technology: The Case of Korea*, London: Croom Helm.

Enos, J.L. (1991), *The Creation of Technological Capability in Developing Countries*, New York: Pinter Publishers.

Equipmentworld Magazine (4 March 2005), equipmentworld.com.

Fialkowski, K. (1990), 'Software industry in developing countries: the possibilities', *Information Technology for Development*, **5** (2), 187–94.

Florida, Richard and Donald F. Smith Jr. (1993), 'Venture capital formation, investment, and regional industrialization', *Annals of the Association of American Geographers*, **83** (3), 434–51.

Folta, T.B. and K.D. Miller (2002), 'Real options in equity partnerships', *Strategic Management Journal*, **23** (1), 77–88, Research Notes and Commentaries.

Foucault, Michel (1980), *Power/Knowledge: Selected Interviews and Other Writings 1970–1977*, New York: Pantheon Books.

Gaio, F.J. (1989), 'The Development of Computer Software Technological Capabilities in Developing Countries: A Case Study of Brazil', DPhil thesis: University of Sussex.

Galbraith, J. (1973), *Designing Complex Organizations*, Reading, MA: Addison-Wesley.

Geertz, C. (1978), 'The bazaar economy: information and search in peasant marketing', *American Economic Review*, **68** (2), 28–32.

Georgopoulos, Basil S. and Floyd C. Mann (1962), *The Community General Hospital*, New York: Macmillan.

Ghose, T.K. and V.S. Bisaria (2000), 'Development of biotechnology in India', *Advances in Biochemical Engineering/Biotechnology*, **69**, 87–124.

Ghose, T.K. (1998), *Technorama* (a supplement to the *Journal of Institution of Engineers, India*), **47**, 23.

Ghoshal, S. and C.A. Bartlett (1990), 'The multinational corporation as an interorganizational network', *Academy of Management Review*, **15** (4), 603–25.

Glaser, B. and A. Strauss (1967), *The Discovery of Grounded Theory: Strategies for Qualitative Research*, Chicago: Aldine Publishing Company.

Goffman, Erving (1967), *Interaction Ritual*, New York: Pantheon Books.

Gompers, Paul and Josh Lerner (2001), 'The venture capital revolution', *Journal of Economic Perspectives*, **15** (2),145–68.

Government of India (2006), Automotive Mission Plan: 2006–2016, Department of Heavy Industry, http://dhinic.in/

Government of India (c. 2006), *Indian Mining and Construction Equipment Industry*, Department of Heavy Industry, Government of India.

Grant, R. (1996), 'Toward a knowledge-based theory of the firm', *Strategic Management Journal*, **17**, Special Issue: Knowledge and the Firm, Winter, 109–22.

Guillen, Mauro (2005), *The Rise of Spanish Multinationals*, New York: Cambridge University Press.

Gulati, Ranjay and Harbir Singh (1998), 'The architecture of cooperation: managing coordination costs and appropriation concerns in strategic alliances', *Administrative Science Quarterly*, **43** (4), 781–814.

Hall, G.R. and R.E. Johnson (1970), 'Transfers of United States Aerospace Technology to Japan', in Raymond Vernon (ed.), *The Technology Factor in International Trade*, New York: Columbia University Press.

Hamel, Gary and C.K. Prahalad (1989), 'Strategic intent', *Harvard Business Review*, **67** (3), May–June, 63–77.

Hannan, Michael T. and John Freeman (1977), 'The population ecology of organizations', *American Journal of Sociology*, **82** (1977), 929–64.

Heeks, Richard (1996), *India's Software Industry: State Policy, Liberalization and Industrial Development*, Thousand Oaks, CA: Sage Publications.

Holland, John H. (1998), *Emergence: From Chaos to Order*, New York: Basic Books.

Holland, John H., Keith J. Holyoak, Richard E. Nisbett, and Paul R. Thagard (1986), *Induction*, Cambridge, MA: MIT Press.

Hufbauer, G. (1966), *Synthetic Materials and the Theory of International Trade*, Cambridge, MA: Harvard University Press, pp. 88–91.

Hughes-Cromwick, Ellen (2003), *Global Auto Industry Trends*, Detroit, MI: Federal Reserve Bank of Chicago, http://www.chicagofed.org/news_and_conferences/conferences_and_events/files/geography_of_auto_production_global_auto_industry_ trends.pdf.

Hymer, Stephen ([1960] 1976), *The International Operations of National Firms: A Study of Direct Foreign Investment*, Cambridge, MA: The MIT Press.

ICFAI (2005), *Innovations in the Biotech and Pharma Industries in India*, Case Code BREP014, ICFAI, India, http://icur.icfai.org/casestudies/catalogue/Innovation/BREP014.htm.

Imai, K., Ikujuro, Nonaka and H. Takeuchi (1985), 'Managing the new product development process: How Japanese companies learn and unlearn', in K. Clark, R. Hayes and C. Lorenz (eds), *The Uneasy Alliance*, Boston, MA: Harvard Business School.

Kaplinsky, R. (1987), *Micro-electronics and Employment Revisited: A Review*, Geneva: ILO.

Katz, Jorge M. (1985), 'Domestic Technological Innovations and Dynamic Comparative Advantages: Further Reflections on a Comparative Case-study Program', in N. Rosenberg and C. Frischtak (eds), *International Technology Transfer: Concepts, Measures and Comparisons*, New York: Praeger, pp. 127–166.

Kindleberger, C.P. (1969), *American Business Abroad*, New Haven, CT: Yale University Press.

Knickerbocker, Frederick (1973). *Oligopolistic Reaction and Multinational Enterprise*, Boston: Division of Research, Graduate School of Business Administration, Harvard University.

Kogut, Bruce (1988), 'Joint ventures: theoretical and empirical perspectives', *Strategic Management Journal*, **9** (4), 319–32.

Kogut, Bruce (1989), 'The stability of joint ventures: reciprocity and competitive rivalry', *Journal of Industrial Economics*, **38** (2), 183–98.

Kogut, Bruce (1991), 'Joint ventures and the option to expand and acquire', *Management Science*, **37** (1), 19–33.

Kogut, Bruce (2000), 'The network as knowledge: generative rules and the emergence of structure', *Strategic Management Journal*, **21** (3), 405–25.

Kogut, Bruce and Nalin Kulatilaka (1994), 'Operating flexibility, global manufacturing and the option value of a multinational network', *Management Science*, **40** (1), January, 123–39.

Kogut, B. and N. Kulatilaka (2001), 'Capabilities as real options', *Organization Science*, Nov/Dec, **12** (6), 744.

Kogut, Bruce and Udo Zander (1993), 'Knowledge of the firm and the evolutionary theory of the multinational corporation', *Journal of International Business Studies*, Fourth Quarter, **24** (4), 625–46.

Kogut, Bruce and Udo Zander (1995), 'Knowledge, Market Failure and the Multinational Enterprise: A Reply', *Journal of International Business Studies*, **26** (2), 417–26.

Kogut, Bruce and Udo Zander (1996), 'What firms do? Coordination, identity, and learning', *Organization Science*, **7** (5), 502–18.

Kramer, R. (2000), 'Trust and Distrust in Organizations: Emerging Perspectives, Enduring Questions', Paper presented at the Organization Science Winter Conference, Colorado, February.

Krugman, Paul (1991), 'Increasing returns and economic geography', *The Journal of Political Economy*, **99** (3), 483–99.

Lala, R.M. (1981), *The Creation of Wealth*, Mumbai, India: IBH Publishing Company.

Lall, Sanjaya (1985), 'Trade in Technology by a Slowly Industrializing Country: India', in N. Rosenberg and C. Frischtak (eds), *International Technology Transfer: Concepts, Measures and Comparisons*, New York: Praeger.

Lall, Sanjaya (1987), *Learning to Industrialize: the Acquisition of Technological Capability by India*, Basingstoke, UK: Macmillan Press.

Lave, J. and Wenger, E. (1991), *Situated Learning*, Cambridge, MA, Cambridge University Press.

Lawrence, P.R. and Jay Lorsch (1969), *Organization and Environment*, Homewood, IL: Richard D. Irwin.

Leonard-Barton, D. (1988), 'Implementation as mutual adaptation of technology and organization', *Research Policy*, **17** (5), 251–67.

Leonard-Barton, D. (1990), 'A dual methodology for case studies: synergistic use of a longitudinal single site with replicated multiple sites', *Organization Science*, **1** (3), 248–66.

Levi-Strauss, Claude ([1962] 1966), *The Savage Mind*, Chicago, IL: University of Chicago Press.

Lewis, W. Arthur (1970), *The Theory of Economic Growth*, New York: Harper Torchbooks.

Licht, Walter (1995), *Industrializing America: The Nineteenth Century*, Baltimore, MD: Johns Hopkins University Press.

Mansfield, Edwin (1988), 'The speed and cost of industrial innovation in Japan and the United States: external vs. internal technology', *Management Science*, **34** (10), 1157–68.

Mansfield, Edwin and Anthony Romeo (1980), 'Technology transfer to overseas subsidiaries by US-based firms', *Quarterly Journal of Economics*, **95** (4), December, 737–50.

Mansfield, Edwin, Anthony Romeo, Mark Schwartz, David Teece, Samuel Wagner and Peter Brach (1983), 'New findings in technology transfer, productivity, and development', *Research Management*, **26** (2), March-April, 11–20.

Maurice, Marc, Francois Sellier and Jean-Jacques Silvestre (1986), *The Social Foundations of Industrial Power: A Comparison of France and Germany*, Cambridge, MA: MIT Press.

McClelland, D.C. (1959), *The Achieving Society*, New York: Free Press.

Meadows, A.J. et al. (1987), *Dictionary of Computing and Information Technology* (3rd edition), London: Kogan Page.

McGrath, R. and Nerkar, A. (2004), 'Real Options Reasoning and a New Look at the R&D Investment Strategies of Pharmaceutical Firms', *Strategic Management Journal*, **25**, 1, 1–21.

Metiu, A. (2006), 'Owning the code: status closure in distributed groups', *Organization Science*, **17** (4), 418–35.

Miles, M.B. and A.M. Huberman ([1984] 1994), *Qualitative Data Analysis*, Newbury Park, CA: Sage Publications.

Mitchell, G.R and W.F. Hamilton (1988), 'Managing R&D as a strategic option', *Research-Technology Management*, **31** (3), May-June, 15–22.

Murphy, Kevin M., Andrei Schleifer and Robert W. Vishny (1989), 'Industrialization and the Big Push', *Journal of Political Economy*, **97** (5), 1003–26.

NASSCOM (2006), *Strategic Review 2006: The IT Industry in India*, New Delhi, India: NASSCOM (National Association of Software and Service Companies).

NASSCOM-McKinsey Report (2002), *Strategies to Achieve the Indian IT Industry's Aspiration*, New Delhi, India: NASSCOM.

Nelson, R.R., (1960), 'Growth models and the escape from Low-level equilibrium trap: The case of Japan', *Economic Development and Culture Change*, **8** (4), 378–388.

Nelson, R.R. (1993), *National Innovation Systems*, New York: Oxford University Press.

Nelson, Richard R. and Howard Pack (1999), 'The Asian miracle and modern growth theory', *The Economic Journal*, **109** (457), 416–36.

Nelson, R. and S. Winter (1982), *An Evolutionary Theory of Economic Change*, Cambridge, MA: Harvard University Press.

Nightingale, P. (2000), 'Economies of scale in experimentation: knowledge and technology in pharmaceutical R&D', *Industrial and Corporate Change*, **9** (2), 315–59.

Noe, Thomas H. and Geoffrey Parker (2005), 'Winner take all: competition, strategy, and the structure of returns in the internet economy', *Journal of Economics & Management Strategy*, **14** (1), 141–64.

Nonaka, I. (1994), 'A dynamic theory of knowledge creation', *Organization Science*, **5** (1), February, 14–37.

OECD (2001), *Biotechnology Statistics in OECD Member Countries: Compendium of National Statistics*, Bridgitte van Beuzekom. Available at: http://www.olis.oecd.org/olis/2001doc.nsf/43bb6130e5e86e5fc12569fa005d004c/c1256985004c66e3c1256ac600350f21/$FILE/JT00112476.PDF. Accessed 11 October 2007.

OECD (2002), *OECD Information Technology Outlook*. Available at: http://www.oecd.org/dataoecd/63/60/1933354.pdf. Accessed: 11 October 2007.

OECD (2005).

OECD (2007), 'OECD Steel Committee sees favourable outlook for steel demand: the global steel industry faces important challenges'; 62nd meeting of the Steel Committee, Istanbul, Turkey, 17–18 May, http://www.oecd.org/document/49/0,3343,en_2649_34221_38622129_1_1_1_1,00.html

Oliver, A.L. (1994), 'In Between Markets and Hierarchies – Networking Through the Life Cycle of New Biotechnology Firms', University of California Los Angeles, ISSR Working Papers in the Social Sciences, 1994–95, **6** (6).

Oviatt, B.M. and P.P. McDougall (1994), 'Toward a theory of international new ventures', *Journal of International Business Studies*, **25**, 1, 45–64.

Pascale, R.J. (1984), 'Perspectives on strategy: The real story behind Honda's success', *California Management Review*, XXIV, 3, 47–72.

Permutter, H.V. (1969), 'The Tortuous Evolution of the Multinational Corporation, *Columbia Journal of World Business*, January–February, 8–18.

Permutter, H.V. (1991), 'On the Rocky Road to the First Global Civilization', *Human Relations*, **44** (9), 897–920.

Piore, Michael J. and Charles F. Sabel (1984), *The Second Industrial Divide: Possibilities for Prosperity*, New York: Basic Books.

Pisano, G. (2000), 'In Search of Dynamic Capabilities: The Origins of R&D Competence in Biopharmaceuticals', in G. Dosi, R.R. Nelson and S.G. Winter (eds), *The Nature and Dynamics of Organizational Capabilities*, New York: Oxford University Press, pp. 129–54.

Popper, Karl (1994), *Knowledge and the Body–Mind Problem*, New York: Routledge.

Porter, Michael E. ([1980] 1998), *Competitive Strategy: Techniques for Analyzing Industries and Competitors*, New York: Free Press.

Porter, Michael E. (1990), *The competitive advantage of nations*, New York: Free Press.

Powell, W., K. Koput and L. Smith-Doerr (1996), 'Interorganizational collaboration and the locus of innovation: networks of learning in biotechnology', *Administrative Science Quarterly*, **41** (1), 116–45.

Prahalad, C.K. and Gary Hamel (1990), 'The core competence of the corporation', *Harvard Business Review*, May-June, HBS Reprint 90311.

Pringle, J.W.S. (1951), 'On the parallel between learning and evolution', *Behavior*, **3**, 174–215.

Raina, Dhruv and S. Irfan Habib (1996), 'The moral legitimation of modern science: Bhadralok reflections on theories of evolution', *Social Studies of Science*, **26** (1), 9–42.

Reddy, N. Mohan and Liming Zhao (1990), 'International technology transfer: a review', *Research Policy*, **19** (4), 285–307.

Rivkin, Jan W. (2000), 'Imitation of complex strategies', *Management Science*, **46** (6), 824–44.

Rogers, Everett M. ([1962] 1995), *Diffusion of Innovations*, New York: Free Press.

Romer, Paul M. (1986), 'Increasing returns and long-run growth', *The Journal of Political Economy*, **94** (5), 1002–37.

Rorty, Richard (1961), 'Pragmatism, categories and language', *Philosophical Review*, **70** (2), 197–223.

Rosenstein-Rodan, P.N. (1943), 'Problems of industrialization of Eastern and South-Eastern Europe', The *Economic Journal*, **53** (210/211), 202–11.

Rosenstein-Rodan, P.N. (1944), 'The international development of economically backward areas', *International Affairs*, **20** (2), 157–65.

Ross, L. (1977), 'The intuitive psychologist and his shortcomings: Distortions in the attribution process', in L. Berkowitz (ed.), *Advances in experimental social psychology (vol. 10)*, New York: Academic Press.

Salganik, Matthew J., Peter Sheridan Dodds, and Duncan J. Watts (2006), 'Experimental study of inequality and unpredictability in an artificial cultural market', *Science*, **311** (5762), 854–6.

Sarkar, S. (2002), *Beyond Nationalist Frames: Relocating Postmodernism, Hindutva, History*, New Delhi, India: Permanent Black, pp. 10–37.

Saxenian, Annalee (1994), *Regional Advantage: Culture and Competition in Silicon Valley and Route 128*, Cambridge, MA, Harvard University Press.

Schware, R. (1992), 'Software industry entry strategies for developing countries', *World Development*, **20** (2), 143–64.

Schwartz, Eduardo S. (2003), 'Patents and R&D as Real Options', NBER Working Paper No. w.10114, November.

Scott-Kemmis, D. and Martin Bell (1988), 'Technological Dynamism and Technological Content of Collaboration: Are Indian Firms Missing Opportunities', in Ashok V. Desai (ed.), *Technology Absorption in Indian Industry*, New Delhi, India: Wiley Eastern Limited.

Selten, R. (2001), 'What is bounded rationality?', in G. Gigerenzer and R. Selten (eds), *Bounded Rationality: The Adaptive Toolbox*, MIT Press, Cambridge, MA, pp. 13–36.

Senge, Peter M. (1994), *The Fifth Discipline: The Art and Practice of the Learning Organization*, New York: Doubleday/Currency.

Simmel, George (1896), 'Superiority and subordination as subject-matter of sociology II', *American Journal of Sociology*, **2** (3), 392–415.

Simon, H.A. (1956), 'Rational choice and the structure of environments', *Psychological Review*, **63** (1956), 129–38.

Simon, H.A. (1965), 'The architecture of complexity', *General Systems*, **10** (1965), 63–73.

Simon, H.A. (2000), 'Near decomposability and the speed of evolution', *Industrial and Corporate Change*, **11** (3), 587–99.

Sorenson, Olav and Toby E. Stuart (2001), 'Syndication networks and the spatial distribution of venture capital investments', *The American Journal of Sociology*, **106** (6), 1546–88.

Sridharan, E. (1989), 'World trends and India's software products', *DataQuest*, December, 53–6.

Srivastava, M.P. (1995), *Development and Planning of Biotechnology Manpower in India*, New Delhi, India: Agricole Publ. Co.

SSKI (2006), *India Research: Auto Components Sector Report*, Mumbai, India: SSKI Financial Services Group.

Stark, D. (2001), 'Ambiguous assets for uncertain environments: heterarchy in postsocialist firms', in Paul DiMaggio (ed.), *The Twenty-first-century Firm: Changing Economic Organization in International Perspective*, Princeton, NJ: Princeton University Press, pp. 69–104.

Stiglitz, Joseph E. (1989), 'Markets, market failures, and development', *American Economic Review*, **79** (2), Papers and Proceedings of the Hundred and First Annual Meeting of the American Economic Association, 197–203.

Stiglitz, Joseph E. (2004), 'Information and the change in the paradigm in economics, Part 2', *The American Economist*, **48** (1), 17–49.

Strauss, A and J. Corbin (1990), *Basics of Qualitative Research: Grounded Theory Procedures and Techniques*, Newbury Park, CA: Sage.

Surie, G. (1996), 'The Creation of Organizational Capabilities through International Transfers of Technology', PhD Dissertation, Wharton School, University of Pennsylvania.

Surie, G. and Harbir Singh (2004), 'Boundary Extensions in Cross-border Innovation via Real Options Heuristics', The Mack Center for Technological Innovation, The Wharton School, presented at the Academy of Management Conference, New Orleans, August, 2004.

Surie, G., R. McGrath and Ian C. MacMillan (2003), 'Options heuristics, knowledge and the timing of technology investments', Paper presented at the Academy of Management, Seattle, August.

Sutton, John (2004), *The Auto-component Supply Chain in China and India – A Benchmarking Study*, London: London School of Economics and Political Science.

Taylor, Frederick Winslow (1911), *The principles of scientific management*, New York, London: Harper and Brothers.

Teece, David (1977), 'Technology transfer by multinational firms: the resource cost of transferring technological know-how', *The Economic Journal*, **87** (June), 242–61.

Thompson, J.D. (1967), *Organizations in Action*, New York: McGraw-Hill.

Todaro, Michael P. (1985), *Economic Development in the Third World*, New York: Longman, Inc.

Tversky, A. and D. Kahneman (1986), 'Rational choice and the framing of decisions', *Journal of Business*, **59** (4), Part 2: The Behavioral Foundations of Economic Theory, S251–S278.

Tyre, M. and Oscar Hauptman (1992), 'Effectiveness of organizational responses to technological change in the production process', *Organization Science*, **3** (3), 301–15.

UNCTAD (2005), *World Investment Report: Transnational Corporations and the Internationalization of R&D*, New York and Geneva: United Nations.

Ungson, G.R., R.M. Steers and S. Park (1997), *Korean Enterprise: The Quest for Globalization*, Boston, MA: Harvard Business School Press.

US Department of Labor (2000–2003), *Occupation Outlook Handbook*.

Van de Ven, Andrew and Diane L. Ferry (1980), *Measuring and Assessing Organizations*, New York: John Wiley and Sons.

Vernon, R. (1966), 'International investment and international trade in the product cycle', *Quarterly Journal of Economics*, **80** (2), 190–207.

Vernon, R. ([1966] 1979), 'The product cycle hypothesis in a new international environment', *Oxford Bulletin of Economics and Statistics*, **41** (4) (November), 255–67.

Weber, M. (1930), *The Protestant Ethic and Spirit of Capitalism*, London: George Allen & Unwin.

Weber, M. (1946), *From Max Weber: Essays in Sociology*, translated, edited and with an introduction by H.H. Gerth and C. Wright Mills, New York: Oxford University Press.

Weick, K.E. (2007), 'The generative properties of richness', *Academy of Management Journal*, **50** (1), 14–19.

Wenger, Etienne (1998), *Communities-of-practice: Learning, Meaning and Identity*, New York: Cambridge University Press.

Westphal, Larry E., Linsu Kim and Carl J. Dahlman (1985), 'Reflections on the Republic of Korea's Acquisition of Technological Capability', in Nathan Rosenberg and Claudio Frischtak (eds), *International Technology Transfer: Concepts, Measures and Comparisons*, New York: Praeger.

Williamson, O. (1975), *Markets and Hierarchies*, New York: Free Press.

Winter, Sidney G. (1987), 'Knowledge and competence as strategic assets', in David J. Teece (ed.), *The Competitive Challenge: Strategies for Industrial Innovation and Renewal*, New York: Harper and Row.

Winter, S.G. and G. Szulanski (2001), 'Replication as strategy', *Organization Science*, **12** (6), 730–93.

World Bank (1992), *India: An Information Technology Development Strategy*, Washington DC: World Bank.

Yin, Richard K. ([1989] 1994), *Case Study Research: Design and Methods*, reprinted 1994, Newbury Park, CA: Sage Publications.

Young, Allyn A. (1928), 'Increasing returns and economic progress', *The Economic Journal*, **38** (152), 527–42.

Zbaracki, M. (1998), 'The rhetoric and reality of total quality management', *Administrative Science Quarterly*, **43** (3), 602–36.

Index